To Sally,

Dancing and Romancing
with *PELE*

Here's what you will do and see
at Volcano

Aloha

Taulan Marie

My Island Publishing
P.O. Box 100
Volcano, HI 96785

Produced by:
Belknap Publishing & Design
P.O. Box 22387
Honolulu, HI 96823

Photographs © Gordon Morse
Additional photograph and art credits on page 102.

Maps © Buzz Belknap

ISBN 0-885129-12-1

Front Cover: The 1960 eruption that covered Kapoho.
Back: Light streams through *'ohi'a lehua* trees,
cornerstone of the indigenous flora that flourishes in the
rich volcanic soil of Kīlauea's summit.

PRINTED IN CHINA THROUGH COLORCRAFT LTD.,
HONG KONG

*1955 lava flow enters ocean
along Puna coast.*

Dancing and Romancing
with PELE

A Story Guide® to Kīlauea Volcano
by Gordon Morse

photographs by the author
maps by Buzz Belknap

My Island Publishing
myislandpublishing.com

Preface

Most books about Kīlauea Volcano put heavy emphasis on its erupting phases. Exciting perhaps, but not what this volcano is all about.

The two-mile-in-diameter caldera, with its pigtail of pit craters topped by heavily forested shoulders, stretches out to an ever-changing black sand shoreline. This is the expanse of quietly dynamic terrain that most visitors see and try to capture on film, or in the mind's eye, often in only a few days. Should visitors reach Kīlauea when an eruption is ongoing and safely accessible to the public, then that may become the sight of a lifetime.

I live on top of Kīlauea Volcano. I have experienced every eruption on the Big Island since 1935. I, too, thrill to the periodic violent action, the glow and shudder of Pele, the fickle goddess of Hawaiian volcanoes. But I also have experienced the quiet, precious side of Pele. You can say I have walked in her hot and cold footsteps.

> *I have felt her many times in spirit.*
> *I have dreamed and fantasized about her.*
> *I met her once in person.*
> *Come with me now, and I'll tell you all about it.*

Gordon Morse
Volcano, Hawaii 2005

Mauna Ulu tree molds, left;
Magma fountains, Kapoho, Puna, right

Dedication

These adventures are dedicated to those,
alive or of late, who shared these events.
True friends all.

Dr. Gordon MacDonald   Robert Wenkam
Reginald Ho   Dr. Jerry Eaton
Arthur Lyman   Slim Holt
Myron Isherwood   Allen Chang
Bill Stearns   Fred Rackle

contents

c o n t e n t s

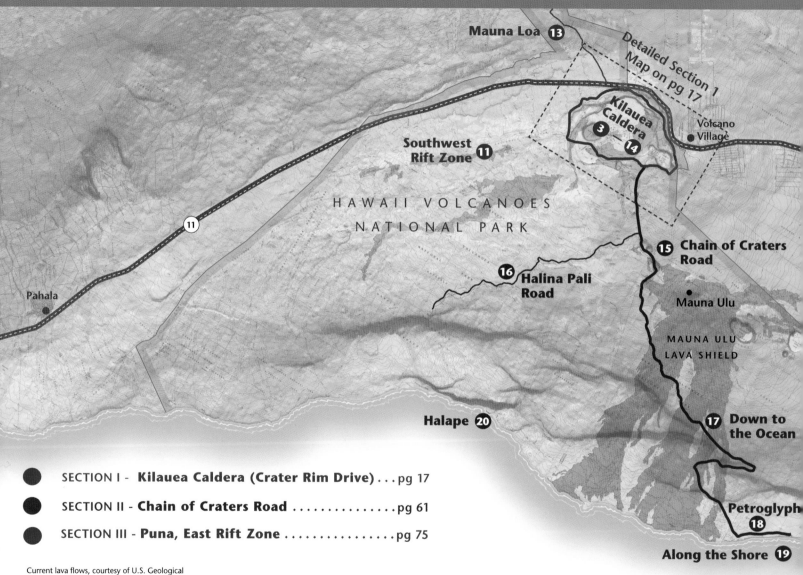

KILAUEA VOLCANO

Mauna Loa (13)

Detailed Section 1 Map on pg 17

Kilauea Caldera (3) — (14)

Volcano Village

Southwest Rift Zone (11)

(11)

HAWAII VOLCANOES NATIONAL PARK

Pahala

(15) Chain of Craters Road

(16) Halina Pali Road

Mauna Ulu

MAUNA ULU LAVA SHIELD

Halape (20)

(17) Down to the Ocean

Petroglyph (18)

Along the Shore (19)

Current lava flows, courtesy of U.S. Geological Survey, Hawaiian Volcano Observatory.

Map © Buzz Belknap 2005

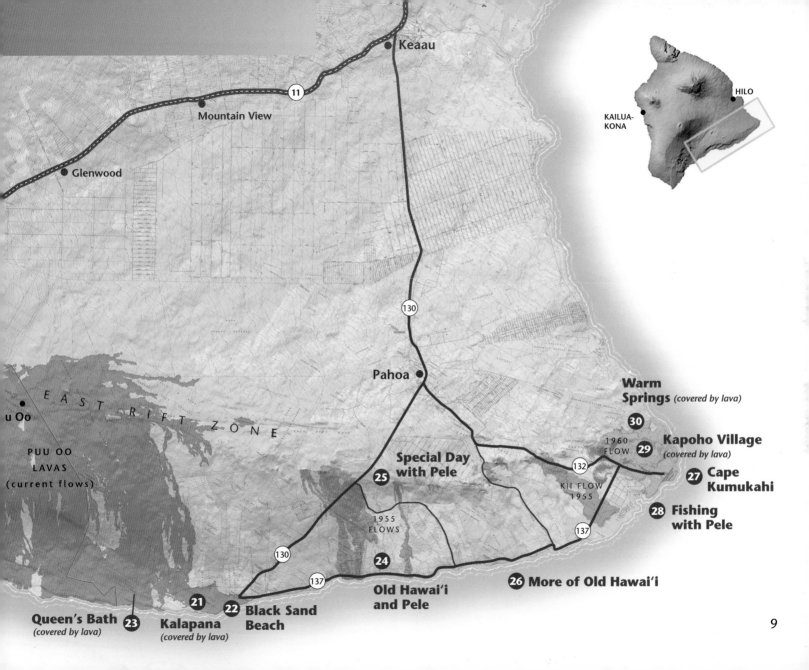

Keaau

Mountain View

Glenwood

11

Pahoa

130

KAILUA-KONA

HILO

E A S T   R I F T   Z O N E

u Oo

PUU OO
LAVAS
(current flows)

Warm Springs *(covered by lava)*

30

1960
FLOW

29 Kapoho Village
(covered by lava)

27 Cape
Kumukahi

132

KII FLOW
1955

28 Fishing
with Pele

25 Special Day
with Pele

1955
FLOWS

26 More of Old Hawai'i

137

130

24

Old Hawai'i
and Pele

137

21

22 Black Sand
Beach

Kalapana
(covered by lava)

Queen's Bath
(covered by lava) **23**

9

Useful Tips Before You Begin Exploring

ALLOW TIME ENOUGH TO SEE THE VOLCANO The highlights of Kīlauea take at least two days to see. Visitors who do in-depth exploring stay from four to five days, and return to do more. Visitors who come on a tour bus, or in a U-drive car for a few hours, are to be pitied. They see little, and experience less of what could be their most meaningful Hawai'i experience.

BY AUTO AND ON FOOT By auto and on foot there's an area greater than the island of O'ahu to explore. Come to think of it, the mountain called Kīlauea, plus all of Hawai'i Volcanoes National Park, which extends from the ocean to the summit of Mauna Loa, is larger than the islands of O'ahu and Lāna'i islands combined.

SECTIONS For your convenience in planning we've divided this book into the following three sections:

Section I- The Summit and Crater Rim Drive Most visitors begin with this 11-mile drive which is a good introduction to what the volcano has to offer. You'll see Kīlauea's caldera with its steaming pots, sulfur banks, craters, lava tubes, fern forest, lava flows, ash-devastated terrain, and natural botanical areas; not to mention man-made goodies like Jagger Museum, the information headquarters, an arts and craft center, a golf course, winery, military rest and recreation camp, and the village of Volcano.

Section II - Chain of Craters Road Going down the east side of the mountain, there's a chain of pit craters with lava tree molds—a landscape of frozen rock with a vista overlooking the entire southeastern side of the island. Below that there's an outdoor studio of Hawaiian rock art, the ocean front with newly-created land, black sand beaches, and historic ruins.

Section III - Puna, The East Rift Zone To the east stretches a tropical jungle with 200-pound ferns perched high on tree limbs, rural Hawai'i as it was years ago, more black sand beaches, the most peaceful picnic parks on the island, and a town that was overrun by lava. All of this is within a tropical flower and papaya-growing district. I haven't even mentioned the walking trails— some that can be enjoyed in an hour, others that take one or two days to traverse. Then, far out on the fringes where backpackers roam, there's a three-day adventure to

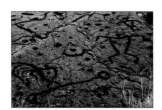   

the summit of 13,677-foot Mauna Loa with overnights in rustic cabins. Finally there is the Ka'ū acid-rain desert.

A GUIDE TO OTHER ESSENTIALS Here's a roundup on facilities available in and around the village of Volcano that will enhance your exploration:

Accommodations Accommodations are limited, so plan ahead with reservations. The Volcano House Hotel has 38 rooms. Volcano Village has a Lodge, several Inns, and the area is famed for its bed-and-breakfast and vacation homes. If you are retired from the military, or on active duty, there's Kīlauea Military Camp. For campers, the National Park offers ten cabins and an excellent tenting area. More primitive camping and picnic parks are located elsewhere.

Arts and Crafts Three wonderful stores: The Volcano Art Center beyond the Kīlauea Visitors Center; Gift Shop atop the Hardware Store; The Quilt Shop by the General Store.

Nighttime Activities There are virtually none, unless an eruption is going on. Then that's the best of all shows. Otherwise, eruption movies are shown nightly at the Volcano House Hotel. The hotel bar is open until midnight and sometimes features a live musical group.

Check the bulletin boards at the Volcano Art Center and National Park headquarters. Several nights a month, usually on Tuesdays, there are visiting lecturers and cultural programs.

Restaurants The Volcano House dining room offers breakfast, lunch and dinner. Kīlauea Military Camp (which has gone public in many respects) offers meals. Volcano Golf Club restaurant serves lunch. Traditional Italian Pizza is next to the Volcano store, Thai food is in the Hardware Store, Lava Rock Cafe is behind the General Store, and the Kīlauea Lodge Restaurant features dinner. There are two stores in Volcano Village. Hilo, with lots of hotel rooms and splendid restaurants, is only 30 miles away on an excellent highway.

Volcano Village Map Use the map on the next page as a guide to things to see and do.

Walks These are described in detail throughout this book, because Pele, the goddess of Hawaiian Volcanoes, also walks these trails daily. Those who choose to walk before 8am and after 4pm will be richly rewarded. Steam, more visible in the cool of the day, gives character, feeling and meaning to the scene. Once the sun heats the atmosphere, the steam issuing from almost every crack diminishes or simply disappears.

Volcano Village

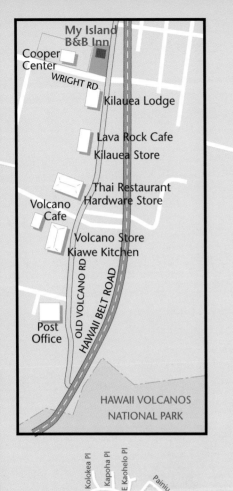

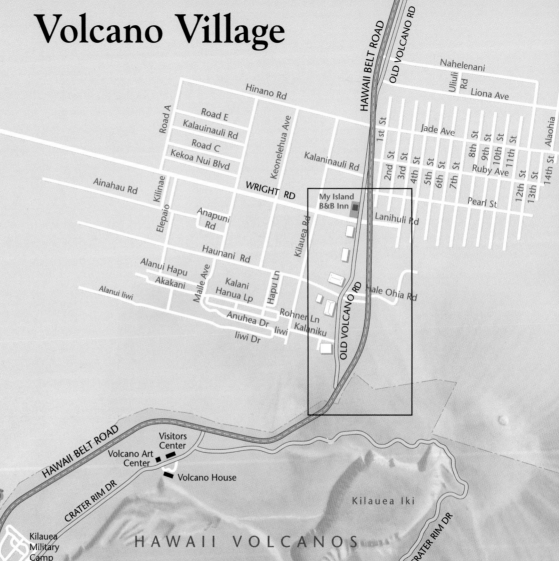

Inset map labels:

My Island B&B Inn
Cooper Center
WRIGHT RD.
Kilauea Lodge
Lava Rock Cafe
Kilauea Store
Thai Restaurant
Hardware Store
Volcano Cafe
Volcano Store
Kiawe Kitchen
OLD VOLCANO RD
HAWAII BELT ROAD
Post Office
HAWAII VOLCANOS NATIONAL PARK

Main map labels:

HAWAII BELT ROAD
OLD VOLCANO RD
Nahelenani
Uliuli Rd
Liona Ave
Hinano Rd
Road A
Road E
Kalauinauli Rd
Road C
Kekoa Nui Blvd
Keonelehua Ave
Kalaninauli Rd
Jade Ave
Ruby Ave
Pearl St
Alaohia
1st St
2nd St
3rd St
4th St
5th St
6th St
7th St
8th St
9th St
10th St
11th St
12th St
13th St
14th St
Ainahau Rd
Kilinae
Elepaio
Anapuni Rd
WRIGHT RD.
Kilauea Rd
My Island B&B Inn
Lanihuli Rd
Haunani Rd
Alanui Hapu
Akakani
Maile Ave
Kalani Hanua Lp
Hapu Ln
Hale Ohia Rd
Alanui Iiwi
Anuhea Dr
Iiwi
Rohner Ln
Kalaniku
Iiwi Dr
OLD VOLCANO RD
Visitors Center
Volcano Art Center
Volcano House
HAWAII BELT ROAD
CRATER RIM DR
Kilauea Iki
CRATER RIM DR
Kolokea Pl
Kapoha Pl
E Kaohelo Pl
Painiu Loop
W Kaohelo Way
Kaakia Pl
Uluhe Way
Konakava Pl
Kilau Way
Pukeawe Pl
Painiu Pl
Puu Oo Volcano Trl
Kanone Pl
Piimauna Dr
Kalehua Pl
Cir
Kanawao Pl
Pukeawe
Kanaio
Kilauea Military Camp
Golf Course Club House
HAWAII VOLCANOS NATIONAL PARK
Kilauea Caldera
Tree Molds
Map © Buzz Belknap 2005

12

An Important First Stop

The logical place to start is at Hawai'i Volcanoes National Park headquarters. The half-hour spent here among the intriguing exhibits, useful books, trail maps, and watching the 20-minute geological movie, will give you enough information to understand what Pele has to offer.

Next door to the headquarters is the hotel that dates from 1877; it has been restored as the Volcano Art Center. Learning about the construction details of the building and seeing its contents both are well worth it.

Across the street through the trees is the Volcano House Hotel built in 1941. Walk through the main lobby and out to the caldera overlook for your initial view of Kīlauea's summit area. Chances are, you are getting your first real glimpse of Kīlauea during banker's hours. Too bad. The beauty and meaning of all this gets lost in the glare of too much sunlight and the swarm of visitors.

Come back to this lookout at sunrise and follow me.

The original Volcano House, built in 1866. Today its replica, the Volcano Art Center, houses an excellent selection of works by contemporary Hawaiian artists and craftspeople.

Welcome to Kīlauea Volcano . . . . .

"Well, you've finally reached the summit of Kīlauea Volcano."

"Where?"

"Where? ...What?"

"Where's Kīlauea Volcano?"

That exchange is typical of those between Big Island residents and first time visitors to the top of Kīlauea and Hawai'i Volcanoes National Park.

The sad fact is that Kīlauea acts like one of the world's most active volcanoes, but it doesn't have the classic image of a lava-spewing mountain. Consider Mauna Loa, which looms over Kīlauea; imposing Mt. Fuji in Japan; Kilimanjaro in Central Africa; or Mt.

Vesuvius in Italy. They have volcanic class. They look like volcanoes.

From any angle, Kīlauea seems shapeless. It resembles a geological afterthought on the slope of massive Mauna Loa Volcano. In fact, Webster's Dictionary erroneously says Kīlauea is a crater on the side of Mauna Loa.

Although Kīlauea's summit is 4,040 feet above sea level, its flanks are so broad that a motorist arriving from the south or east on the Big Island's Route 11 has no feeling of having gained that much altitude.

Suddenly you are just here after turning off the main highway and driving toward the headquarters

. . . . . the sight of a lifetime

building of Hawai'i Volcanoes National Park. Upon arriving there's no hint of a volcano. There are only 'ōhi'a trees shaking their red pompons of lehua blossoms, umbrellas of tree ferns, and a cool, cloud-crying atmosphere.

"So where's Kīlauea Volcano?"

Someone points. You follow directions and behind a red two-story wooden hotel of modest size, or down several cinder trails under the ferns, you come to the edge of a gigantic hole in the ground—Kīlauea Caldera.

"Okay, so where's the action? Where are the blow-torch fountains, the rivers of molten rock, clouds of ash, and Pele's hair covering the forests?"

"Those things could appear at any time," a National Park ranger answers. "Sometimes there is action inside Halema'uma'u crater on the caldera's floor, but more often somewhere along the volcano's east rift, that weakness in the earth that runs from the summit to the Puna shore, and 70 miles out to sea. And if you are lucky," he'll add, "Go right now and see the show. It's the sight of a lifetime."

Whether this mountain is erupting or not, I offer you a highly personalized and completely non-commercial tour of Kīlauea Volcano.

Follow me.

Kilauea Caldera (Crater Rim Drive)

Bird Park

13 **Mauna Loa**

Tree Molds

Golf Course

Kilauea Military Camp

BEGIN HERE
Kilauea Visitors
Center

Wright Rd

Seasons at Kilauea
(Gordon's Garden)

14

Mauna Loa Road

Volcano Art
Center

0

Entrance
Station

VOLCANO
VILLAGE

Old Volcano Road

11

Crater Rim Drive

Sunrise Walk **1**

Volcano House

11

Waldron Ledge
'Earthquake' Walk **2**

12 **Jaggar Museum**

3 **Kilauea Iki Overlook**

4 **Kilauea Iki**
Walk

5 **Thurston**
Lava Tube

K I L A U E A C A L D E R A

Perfect View **7**

HAWAII VOLCANO NATIONAL PARK

10 **Halemaumau**

6 **Devastation**
Trail

Crater Rim Drive

The Sunset Spot

Chain of Craters Road

South West
Rift Zone **11**

9

8 **Keanakakoi**
Crater

K A U D E S E R T

Map © Buzz Belknap 2005

17

Kīlauea Caldera

Standing on the north rim of Kīlauea's caldera, try to imagine this scene as geologists say it was not too long ago, and how it may become again.

Today we view a caldera two by two-and-a-half miles across by about 400 feet deep. A mid-sized U.S. city like Indianapolis could fit inside this depression.

To your left (east) and right (west), the perpendicular walls of the caldera reveal its depth. The lower southern rim across the way is less than 100 feet high, and if this hole in the ground were to fill up with molten lava it would overflow toward the uninhabited, bleak wasteland down the southern flank of this 4,000-foot high volcano. That is why the Volcano House Hotel, National Park headquarters, military rest camp, a golf course and nearby Volcano village of about 3,000 people are on the north rim. This is the "safe" (for now) high, upwind side of Kīlauea.

Wisps of steam issue from cracks across the hardened lava flows that form the desolate floor of the caldera.

Over at the southern corner is a smaller hole in the ground—Halemaʻumaʻu Crater—the modern historical hot spot of this volcano.

That long loaf of mountain you see spanning the skyline to the west is called Mauna Loa. Two hundred years ago the Hawaiians called it "Mawna Roa." Today, historic researchers say the Spanish discoverers of Hawaii 400 years ago called it "Las Mesas."

The scene in and around this caldera can change during one person's lifetime. My grandfather, back in 1880, saw a wholly different picture here. My great-grandchild may not see what I view today. That's geologically exciting when you think of landscapes like Niagara Falls, the Grand Canyon, or even Crater Lake in Oregon, where the only changes over the last few thousand years have been man-made items like roads or hotels. We don't think of major earth shapes changing from generation to generation.

Look to the right at the highest portion of the caldera wall near the Volcano Observatory, an area called Awekahuna Bluff. The last lava flow that spilled over the top of this sheer wall is only 500 years old. A wave of lava rolled over the rim we are standing on when America's east coast was being first colonized. The oldest ash near the bottom of this cliff is a little over 2,000 years old. This means that it is entirely possible that human eyes saw Kīlauea *before* this huge caldera even existed. Egypt's pyramids had already begun to deteriorate. Rome was reshaping the world in its image. The Hebrew prophets were anticipating the coming of

The snows of Mauna Loa and the fires of Kīlauea.

a Savior when volcanic activity was just beginning to build Kīlauea's present caldera. And Columbus had just booked his cruise to the Caribbean when the caldera's wall was completed.

Anthropologists tell us Polynesians migrated here from the south central Pacific (probably from the Marquesas) at least 1,900 years ago. How did these earliest of Hawaii's residents view Kīlauea? Geologists believe the first volcano watchers saw a wholly different caldera. They feel it could have been what they now call the Powers Caldera, which has been covered over with the existing caldera. They stress that back then more of the eruptions were along the rift fault zones, especially in the Puna area, than at Kīlauea's summit.

Sometime after the birth of Christ, the entire top collapsed again. It was not a spectacular all-at-once extravaganza as Hollywood might portray it, but featured some downward movement here and there over a period of years accompanied by earthquakes, eruptions and ash blowouts. The floor was still collapsing in the early 1800s when missionaries first saw the top of this volcano. The Kīlauea caldera they viewed was a vastly different depression from what we see today. For one thing, it was far, far deeper—two to three times what it is now. There were at least five hot-spot pits around the floor. And boy, could they put on a show when the pressure was up!

The first written account of Kīlauea was by the Rev. William Ellis, a British missionary, in 1823. The caldera's floor was about 900 feet deeper than it is today.

Why did it take so long for white man to see and record this volcano? British Captain James Cook sailed past the perimeter of this volcano 46 years earlier, and so did perhaps fifty other sailing ships after him, but the Ellis-led missionary party was the first to explore around the Big Island on foot. Why the delay in someone climbing this modest 4,000-foot mountain to see what was going on?

Perhaps one answer is that Kīlauea wasn't a visually imposing volcano from a distance. Mauna Loa was a far grander mountain with its occasional fire, winter icecap and flows that raced 50 miles to the sea. The early explorers climbed Mauna Loa instead.

At any rate, once someone had written about the ease of walking or riding a horse to view the "everlasting volcanic fires of Pele at Kīlauea" the rush was on. "If you haven't visited Kīlauea you haven't seen Hawaii" was the slogan of the 1800s. More was written about this caldera in the nineteenth century than in the past 100 years. It wasn't the sandalwood trees, the whaling ships, the sugar, or the hula girl that made Hawaii famous, it was Kīlauea caldera. Read Mark Twain's *Letters from the Sandwich Islands* if you don't believe me.

But back to viewing Kīlauea. What you might not know is that this volcano has had two rims: the one we are standing on, and another one 500 feet below the present floor. This lower rim was called "The Black Ledge." This was the catwalk from which all early volcano watchers viewed activity within the caldera during much of the last century.

Kīlauea caldera's greatest depth may have been as much as 1,500 feet. That's the depth estimated by Lieutenant Malden of the HMS Blonde, the U.S. expeditionary voyage that reached Hawai'i in 1825. This would put the level of the floor at about the elevation where the community of Glenwood is today, which is half way down to sea level.

Volcanologists now say that what was seen of Kīlauea's caldera from 1823 to 1900 was a temporary and unusual phase of the volcano. Activity was mainly centered in the caldera, and it was more or less constant. In the depths of the canyon, boiling cauldrons of lava were spitting all over the floor. In 1848 a bubble-like lava dome almost a mile in diameter grew so big it was higher than the outer rim. Then it too, collapsed, leaving spire-shaped walls that somewhat resembled burned-out cathedrals. In time, the summit activity slowly built up the floor to the way we see it in today.

The steam eruption of Halema'uma'u and its collapse between 1919 and 1924 apparently signaled the end of this 100-year phase. Since then Kīlauea has been returning to its normal prehistoric characteristics of eruptions on its flank rift along with only occasional summit activity. There were no more continuous shows inside Kīlauea no matter how often old George Lycurgus, who owned the Volcano House Hotel, threw bottles of gin into the caldera for Pele's happy hour. Since 1955 there have been more and more flank eruptions.

Now as we stand looking over Kīlauea we may wonder what's going to happen next. Will the caldera floor slowly keep building toward the rim? Will there be another collapse? Is such a collapse imminent because of the eruption that has been going on for the past 21 years, where so much erupting material has drained away from under our feet? I wonder if such a collapse of the summit will happen in my lifetime? In a way I hope so, and yet in another way I hope not. I live on the high "safe" side of Pele's home. The skin of land under Volcano Village was created from eruptions in volcanic activity in 1790 when the caldera was in its final stages of collapse. People living here are definitely vulnerable. Perhaps Pele regards us as expendable in the overall plans she is now making.

At any rate, you should take pictures of Kīlauea now so your great-grandchildren can make a comparison of this volcano in 2090 when they come to visit.

Kīlauea caldera in 1841 as seen by Lieutenant Charles Wilkes of the United States Exploring Expedition.

Sunlight filters through old growth ʻōhiʻa lehua trees at Kīlauea.

Exploring Pele's Secret World

Walks of Discovery

The prime time of a Kīlauea day is from the promise of a cool dawn to 8am when the 'ōhi'a forest birds are feeding; and from 5pm to dusk when clouds begin to settle in, and the last glimmer of light bleeds away over Mauna Loa's crescent dome. This is when Kīlauea best displays its treasures.

❶ Sunrise Walk

My sunrise trail is the one mile walk through the steam vents. It's an inexhaustible feast for the senses. Every day is unique. It's as though you are seeing Pele wearing a different muumuu each time. I prize her cool weather attire. She shows off this wardrobe from December through March.

The trail begins at the entrance to the Volcano House driveway across the street from the National Park Headquarters parking lot. It starts cold and hard underfoot on a macadam-paved path. Directly behind me the first yellow shafts of sun filtering through 'ōhi'a paint the air a cream color, and the newly-shingled side of the 1877 Volcano House a glowing tangerine.

A soft pad, pad, pad warns me to move aside. A jogger in gray sweat pants, maroon turtleneck sweater and a brow band that says "I love Hawai'i " hurries past.

"Hi. Out early, huh?"

"Yah. Gotta get in my four miles before breakfast." And off he goes without breaking stride. I expected him to ask, "What are you doing up so early?" But he's gone.

Immediately I enter a thicket of sword ferns jabbing their yellow-green blades skyward, gingers with floppy leaves and fountaining hapu'u ferns. Under this canopy the air suddenly grows warm and sticky. The earth sighs and its white breath rises to linger under leaves as if it is afraid to set out on its own in a wider world.

The walk dips down a series of steps carpeted with moss. Leaning stalks of ginger wave their yellow flowered clusters in my face. The sweet, earthy perfume is pungent. At times I wonder if it should be filtered before breathing. Are our body systems up to taking in this heavy bouquet so early in the day?

Now out into an open, brush-choked meadow. The path here is cinder turning to hard-packed soil. Blobs of red and pink appear within reach—'ōhelo berries that grow and resemble bush blueberries except for their color. I pop a handful into my mouth, expecting a taste from such a luscious, fat, ripe, juicy berry. But it's not there. Insipid, everyone says. Webster says that means "without

'Ōhelo berries… "a distinctive taste of their own"

a distinctive taste of its own." That's debatable. There is a hint of highly diluted cranberries. Perhaps if you cook them with a squeeze of lemon, some diced tart apple, and a spoonful of crushed pineapple, the taste will be enhanced?

Into the 'ōhi'a woods again with apple-green wāwae'iole club moss growing like foot-high Christmas trees.

Suddenly I'm at the very edge of Kīlauea Caldera and a cold, dew-dampened iron rail keeps me from falling 400 feet onto the crater floor. Half of that floor nearest me is a blue shadow since the sun hasn't hurdled the east rim yet, but the far southern plain of the caldera is floodlit a suede tan.

Since the air hasn't quite let go of the night's coolness, every crack in Pele's frozen lava bed issues veils of steam. History says that the two-mile-wide, egg-shaped depression once had five major "hot spots." Beginning just before this century Halema'uma'u overrode the others and covered them 1,000 feet deep in lava flows. But the five cannot be denied their existence. Their locations are exposed by rings of steam this morning.

The trail now leads to visual excitement. First, it skirts the caldera's rim so closely that I quicken my pace. There's graphic evidence of the rim cracking and being in danger of cascading into the caldera. The ridge on my left has already separated from the path by a good three feet. Perhaps if I brace my feet against this ridge and give a shove it will topple. Here on my right is another crack. A good earthquake at this very moment would throw the whole trail and me into the chasm! I hurry.

Now I make a bet with myself. Within the next 500 feet I will see a honeycreeper. Probably the flash of scarlet that is an *'apapane* breakfasting on nectar in the cluster of tiny cups that make up the red lehua blossom. I always win my bet. There he is, a flitting bundle of energy, darting from flower to flower, dipping quickly into his calorie cup while keeping an eye on me. How many flowers do you have to visit for a meal? How quickly does each lehua blossom replenish its nectar?

The forest now thins and I enter a fantasy of swirling cloud. Cracks in the ground cough up steam on all sides. One minute a dense cloud dims my way. Then a breath of breeze catches the whiteness and throws it into the trees. Sunlight, trying to penetrate the thicket, resembles spotlights in a smoky, darkened theater.

This steaming cloud has sound. The audio is hard to pinpoint. One part of it is a drop of water sizzling on a hot stove lid; another is like crickets singing underwater; or water humming in a pan just before it comes to a boil.

There are smells also. The whiff of a soft-boiled egg when you first crack it open. The powder used on athlete's foot. Beet greens cooking. Someone making a stew out of dirt and an old wool blanket.

If you're into the intricate scenes of nature, face the rising sun and look at the slender shoots of *pili* grass. Steam dew decorates the stems with drops that glisten like back-lit soap bubbles. A clump of the grass resembles sparkling upside down icicles.

The fern-covered sides and depths of the lazy steam vents are a world of their own. The growth is luxuriant and luminous with greens and yellows far richer than those outside the pits. The warm bowls and cracks are living ads for the color film which promises "a brighter-than-real-life" color.

This scene soon gives way to a true phenomenon that takes many of these walks over many months to appreciate. The trail goes along the edge of the caldera; the very edge is one continuous steaming curtain.

Almost daily, this curtain will put on a different show. It goes straight up—crooked—wafting back and forth. It blows north. It blows south. Most spectacular of all, it becomes a heavy milky vapor fall that cascades down the 400-foot sheer wall of the caldera to the bottom, a silent Kīlauea Niagara.

"Pound, pound, pound." My jogger is returning. His feet are heavier and I notice a panting before he materializes out of the steam like a ghost come to life.

"Hello again," I say.

"Uh, huh," he responds and waves a hand to acknowledge my presence. Then "thud, thud, thud," he heads for his hotel room, a hot shower and a no-cholesterol breakfast at the Volcano House hotel. I can't jog. It makes my kidneys ache. But I do have to hurry. The sun has risen above Kīlauea's rim now and in another 15 minutes, Pele's secret world will disappear.

I scoot across a field with bald spots beginning to grow silver-gray lichen-like fuzz. Across a highway with its yellow ribbon down the middle and into another field covered chest-high with tangles of *uluhe*, or staghorn ferns. This is miserable stuff to walk through, so no one does. In several years it can reduce a forest to its size by smothering the trees. Here more steam vents tell the fern to limit its territory-grabbing ways. If it doesn't, the steam turns it all into a tangle of aluminum straws.

The trail curves out of this thicket beside a meadow of pili and broomsedge grass. Against the far side is a wall of rock the height of a two-story row of apartment houses. The bank is steaming, its sides glowing in a haze of yellows with reddish-orange highlights. From here it's a picture I've never been able to capture.

The scene I dream of capturing on film here is the sulphur bank steaming, all aglow through the shafts of a wheat field of golden pili grass, with everything in focus like those pictures in *Arizona Highways* magazine. I take three more pictures as I'm in a hopeful mood this morning.

On close inspection, the steaming cliff smells of rotten eggs reminiscent of my high school chemistry lab. The smell is not overpowering, just a nuisance when you're trying to get up close to view the crystals of sulphur clinging to the rocks.

My eye sees more than any film can record. Sulphur colors come in three shades of yellow mixed with pea green or aqua. Some rocks are burnt orange with pink curls like that fancy ice cream with flavor swirls all through it. Other formations the size of a car's steering wheel are a metallic blue with battleship gray trim around the edges.

At this point a couple comes out of the forest walking the steam vent trail the wrong way. The man aims his camera at the sulphur scene and goes "click."

His wife tugs at his sleeve and says, "Hurry up, honey. Let's get out of here." She has heavy blue eye shadow and strings of tiny bells for earrings that dangle to her shoulders. "This place stinks!"

They turn around, hasten their stride, and go back into the forest the way they came, headed for the Volcano House Hotel. They have seen and experienced all that Pele offers at dawn.

My morning walk with Pele is over. The trail goes through another ginger jungle back to where my car is parked.

I'll only pause at one more spot out of respect for history. On a bluff overlooking the area I have just walked is a field mostly gone back to nature with koa trees now 30 feet tall. The National Park Service has built a ceremonial dance platform here. This is the spot where all the volcano's prehistoric leaf shelters, grass huts, crude houses, and finally rustic hotels were situated. The present one was built in 1940 on the very rim of the caldera.

❷ The 'Earthquake Walk'

To explore a section of Kīlauea's rim called "The Earthquake Walk," go left along the trail fronting Volcano House. The trail connects to a section of the former Crater Rim Road that did a dance in November 1983 during Pele's 6.6 magnitude earthquake rhumba. Part of the highway, along with sections of the crater rim, fell into the caldera. The most interesting sight comes near the end of the walk where only a sliver of the highway seems to span thin air.

Hawaiian pili grasses flank the hazy greenish glow of Kīlauea's sulphur banks.

'Ōhia lehua haole. Introduced to Kīlauea in the 1800s from Bolivia, this shrub or small tree produces plump blooms. Sometimes called "powder puff" lehua today, they are treasured in leis for their sturdiness.

Crater Rim Drive

The eleven-mile Crater Rim Drive will give you a good start understanding what Kīlauea is all about. If you are beginning this tour before 2pm, go clockwise. If after 2pm go counterclockwise. Reason: the United States Geological Observatory's Jagger Museum of Volcanology closes at 4:30pm. This is an important part of your tour.

The summit area is best experienced during the morning and early afternoon. Clouds scudding in from the northeast tend to increase and cry on the land at this altitude later in the day.

Leaving the National Park Headquarters area and going clockwise on Crater Rim Drive, the road curves through a forest of 'ōhi'a and hāpu'u ferns typical of the wet areas of Kīlauea.

'Ōhi'a is a hard wood with the density of oak. It has so many varieties that botanists have a hard time classifying them all. The tree may be a bush such as is found in mountain swamps on Kaua'i and the highlands of Tahiti, towering trees as on the windward slope of Kīlauea, or the bent over twisted giants that still live on the wet side of Mauna Kea. The wood was used in the making of idols and gunwales of outrigger canoes in bygone days. Back around 1890, whole forests were cut down when the Santa Fe Railroad let out a contract for 'ōhi'a and koa railroad ties. Today, those of us with fireplaces usually burn two cords of 'ōhi'a over the winter months.

The 'ōhi'a's flower is a cluster of brushes called *lehua*, in colors that run from a salmon pink through various reds, an orange, and even bright yellow. It was the flower widely used in the making of lei in the "olden days." Today it is rarely used, as imported flowers such as plumeria, ginger and vanda orchids are more accessible, hardier, and more fragrant. Also, people shy away from picking the flower as a superstitious old wives' tale surrounds the lehua. The claim is that it will rain if you pick the blossom. Well, not entirely true. The threat evolved from native Hawaiians' belief that the flower should only be picked when one was headed for home after spending the day in an 'ōhi'a forest, because the plucked blossom caused clouds to settle on the land and a person could lose his way. The lehua has been voted the official flower of the Island of Hawai'i, and the tree now enjoys a double name, 'ōhi'a lehua.

Enjoy the hāpu'u fern while you can. It is slowly disappearing from our forests as the world turns drier and wild pigs root into the plant. Furthermore, Hawaiians nearly killed entire forests of hāpu'u in the 1800s when they harvested the fuzz from its stems, baled it like cotton and shipped it to San Francisco where it was used as stuffing for mattresses and pillows. Visitors from other Hawai'ian islands are forever taking plants home to grace their patios. The one thing that will certainly doom the hāpu'u fern is the establishment of a vehicle ferry system between the islands. Then it will be easy for anyone to rape the forests and cart ferns home in pickup trucks. The trunk, when shredded, becomes a medium for growing orchids. I knew this drive when the fern made a canopy over the road. Now it is more like a patio plant in hiding within the 'ōhi'a forest.

❸ Kīlauea Iki Overlook

First stop along this Crater Rim Road is the lookout across Kīlauea Iki Crater (Iki means small). On clear days, photographers thrill at this scene. In one photo they capture this small crater, much of the larger caldera beyond with steaming Halema'uma'u Crater, and the outline of Mauna Loa Volcano. The information about the 1959 eruption on the display board helps you to understand what you view from here, but there are worlds more to this crater.

Motor another few thousand feet and park at the Thurston Lava Tube, then follow me on the single best walk this volcano offers.

Hapu'u fern…"enjoy the hapu'u fern while you can."

4 Kīlauea Iki Walk

Friends come to visit Kīlauea for a weekend, learn all there is to do and see, and say, "We didn't really know. There is so much to do. We only have this short time. What's the one meaningful walk that will give us a feeling as to what this volcano is all about?"

"Walk Kīlauea Iki," I respond. "It's the most satisfying experience around here. You can do the 3.4 miles in about two hours. The best time to go is before 8 in the morning because you need the cooler part of the day. As the atmosphere heats up, steam vapor disappears and you've lost some of the best effects of the walk. Also, it is unpleasant walking on the frying-pan floor of an active volcanic crater when there's heat below and above you."

I take this walk right after sunrise once a month. It starts and ends at the Thurston Lava Tube parking lot. Walk counterclockwise. Visitors who don't know the score go clockwise.

The trail begins as a pussycat and skirts the edge of Kīlauea Iki Crater. The views here are for photographers, as one frame can capture the crater, some of Kīlauea Caldera beyond with Halemaʻumaʻu steaming, and the sweep of Mauna Loa.

At this point it helps to know the difference between a crater and a caldera. Both are volcanic holes in the earth. One that's over a mile in diameter is a caldera, one less than that is a crater. That's the way it has been since geologists began writing books. But not today. They changed it to: *a caldera can be any size hole in a volcano's summit, if it is the **only** hole.* We'll regroup later for a chemistry lesson on the crater floor.

Warning! You are approaching the tourist lookout for Kīlauea Iki. That's bad news if there is a tour bus here. For some reason, every male passenger has an urge to water ferns in the ʻōhiʻa forest at this stop. They do it by standing in a line out of view of female passengers who are listening to the bus driver tell about Kīlauea Iki's 1959 eruption. The women are aware of the men's activity and they are envious and restless because they have to wait until the next stop where there are restrooms.

Now we go down a flight of stone steps and then along the entire north rim of the crater. You may hear some of our native forest birds speaking in authoritative tones but you may not see them.

For the last three or four
decades the walk across
the floor of Kīlauea Iki
Crater has been one of the
best outdoor adventures
in the world. The lava
beneath what was a fiery
lake in 1959 has solidified,
but not completely.

The Chinese Ground Orchid

What may catch your attention is a broad-leaved plant that looks suspiciously like skunk cabbage. It's the Chinese ground orchid. The flower is worth a close look. Its petals are brown on the front and pure white behind. The trumpet of the orchid is a muted mixture of lilac and yellow. Why this plant is being ignored by nature writers is a mystery. Four popular booklets on plants of this National Park have been produced and none mentions this orchid. You have to go into the serious botany texts to find it. Then you discover it is one of the oldest orchids on record, introduced here around 1852 from China. Maybe we are ignoring it for political purposes in the interest of world peace.

Lava Fountains, Cinder Cones and Forests

There are areas along the trail here that offer a grand view of the crater. How different it is from when I first explored it in 1943! Back then it had two levels, was twice as deep, and forested on the bottom. It looked very ancient then, but wasn't. Barely 500 years ago eruptions from here created the Thurston Lava Tube. Steams of the molten stuff also covered what today is Volcano Village where I live. Kīlauea Iki's floor did a final collapse in 1868 and gradually became forested.

"Just before Christmas 1959, a 1900-foot fountain helped turn the floor of Kīlauea Iki Crater into a lake of molten lava."

Just before Christmas 1959, the wall directly across from where we're now standing cracked halfway up its face. Over the next month erupting lava filled half this crater. The crumbling hill beyond is the cinder cone constructed by a lava fountain that reached 1900 feet at one point. At the base of the cone is the vent now at the new floor level. Walk with me down to it.

Today Kīlauea Iki resembles a bathtub in which eleven million football players washed off their mud and grime after a rainy day's game. That accounts for the 50-foot-high bathtub ring around the edge. We'll inspect the ring down there, but first we have to get off Kīlauea Iki's rim and drop down to Byron's Ledge. An easy switchback on the trail does it.

Lord Byron's Ledge

During the first half-century of the white man's exploration of Kīlauea, this ledge was a favored camping spot and a platform for those descending to try their luck in the caldera. Lord Anson Byron, cousin of the poet and captain of the *HMS Blonde,* led the first scientific expedition to map Kīlauea. He stayed here— hence the ledge's name. Chiefess Kapiʻolani diminished the fears of the Hawaiians and the power of ancient Hawaiʻian gods from this ledge when she entered the caldera, renounced Pele, and returned unharmed.

The trail now leads to what we call "the back side of the crater." Down the sheer wall we go on a switchback track that periodically cascades into the crater during earthquakes. I hurry down with the same feeling I get when I'm only halfway across a busy thoroughfare and the light turns yellow.

At the lava-crusted bottom the walker is introduced to *ahu,* signposts of piled rocks marking the way through any area that needs it. Look ahead from one ahu to the next to stay on the established path.

One sight here gives pleasure: first growth from seemingly solid rock. Blades of sword ferns, miniature ʻāmaʻumaʻu ferns and ʻōhiʻa trees with their new apricot-colored leaves are gaining a foothold in cracks where wind-blown ash has accumulated.

Watch for Examples of Lava

Here, too, are good examples of *pāhoehoe* and *ʻaʻā* that puzzle those who are new to a volcanic landscape. Surely the smooth, almost glassy pāhoehoe, and the busted up chunks of aʻā that can rip the boots off your feet, are different magma. Nope. Same rock. The difference is that gas trapped in the magma suddenly escaped causing the crumbling situation in what we call ʻaʻā. There's a sign at the throat of the eruption site. It is accurate. If you can't read, then listen. Remember the sound of your grandmother (or great-grandmother) lifting the lid off the old cast iron woodburning stove? Remember the hollow "clang" when she dropped it back on? That same sound echoes with every step you take here. To fool around here is to be very foolish.

Leaving the area you walk across tilting tabletops

of pāhoehoe jumbled like slabs of river ice at spring breakup. These plates were broken and rearranged when the lava lake built higher than the fountain feeding it and the molten stuff drained back into the vent.

Two ridges converge on the trail. On cool, crisp mornings steam lazes all around, blowing a white-out across the path, rising to allow you to glimpse the way, then fogging in again. A black and white world fades in and out of focus as you walk through the gates of hell.

Keep a happy thought while sauntering across the rest of Kīlauea Iki's floor. There's still molten lava —underfoot. The thickness of this lava lake floor is 360 feet. The bottom 120 feet are still fluid.

Alien squares of concrete and ends of pipe on the lava surface appear in the crater's center where scientists drilled to record the rate of cooling. Eight years after the 1959 eruption the crust had cooled to a depth of 81 feet. In 1975 it solidified to 165 feet, and in 1986 to 235 feet.

What look like spider webbing over the lava are cracks, frothing at their lips with yellowish-white or rusty residue. Having flunked chemistry in high school I can only guess it is composed of magnesium sulfates, silica, and thermal oxidation.

Slab-like plates of uplifted pāhoehoe (left front) flank smoother pāhoehoe, with chunky 'a'ā at the rear. The pattern is typical of lava flow visible on Kīlauea Iki's caldera floor and elsewhere around the volcano.

Cooking Lunch on the Crater Floor

Gingerly I lower my hand into one of the cracks and yank it back. Hot! I do this in memory of my four children. As young teenagers they delighted in walking Kīlauea Iki's floor with lunch. Visitors stumping across this moonscape would see the young people seated around one of these cracks poking sticks into its depths.

"What are you doing?"

"Cooking our lunch," one would respond, drawing up his stick with a plump and steaming pink hot dog skewered on the end.

Mark Twain's Breakfast Strawberries

The floor walk ends as you scramble up the bathtub ring created when the floor sank as some of the magma drained back into the vent while the rest cooled. The trail is an easy grade switchbacking into the damp coolness of an ʻōhiʻa and hāpuʻu fern forest.

Crowding the edge of the walk are wild strawberries like those Mark Twain ordered for breakfast in 1866. The aroma is just fine but the sight of the berry is disappointing. They are an Old World variety the size of your little fingernail and pure white when ripe.

Upper left, pāhoehoe in the making—at 2000 degrees Fahrenheit or more, it's strictly a "Don't touch!" item; upper right, miniature sword ferns emerge from a bed of rope pāhoehoe lava.

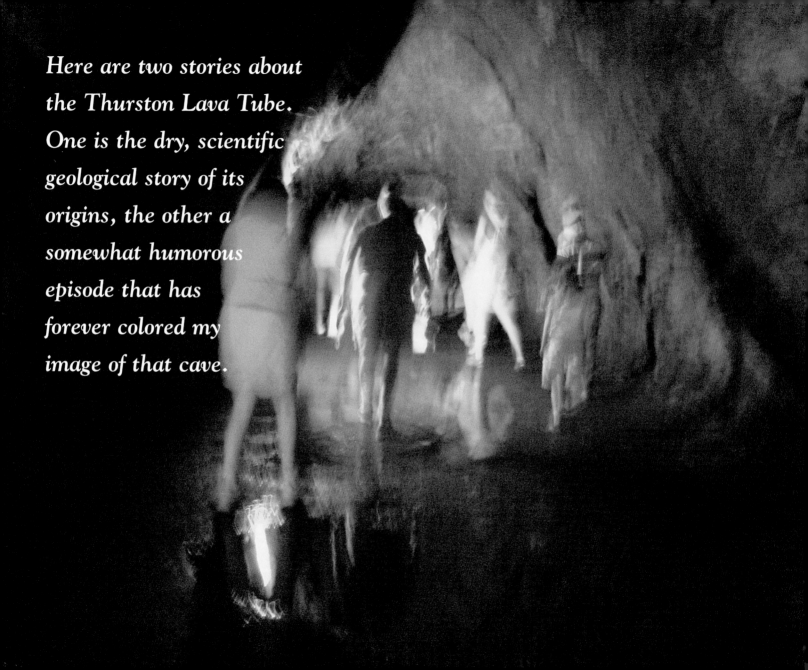

Here are two stories about
the Thurston Lava Tube.
One is the dry, scientific
geological story of its
origins, the other a
somewhat humorous
episode that has
forever colored my
image of that cave.

5 Thurston Lava Tube

The cave-like tube was created by a lava flow from Kīlauea Iki sometime between 1450 and 1700 AD when streams of liquid rock poured the foundation on which Volcano Village was built and laid down a skin on much of the land down towards Hilo. When the dome of Kīlauea Iki collapsed, so did pockets of land around its perimeter. One such pit crater exposed this lava tube, which was not detected until 1913 when Lorrin Thurston, publisher of *The Honolulu Advertiser*, discovered it. He had been a prime mover to get Kīlauea into the roster of National Parks.

In the 1960s there were no lights in the cave. Then National Park safety guidelines for such caves took all the spooky adventure out of it. A paved walk and street lighting were installed.

Today most visitors view this lava tube in the geological sense. But I still remember when my four kids and their friends (sons and daughters of the Park Superintendant and rangers) tried their best to put mystery back into that cave.

Kīlauea Volcano is the playground for youngsters who live around here, but after awhile, even this extravaganza of nature can get boring to active young people. One summer back in the late 1960s the kids dreamed up some excitement. They harvested moss from the surrounding forest and fashioned it into light fluffy balls. They unscrewed several light bulbs in the middle of the lava tube, climbed up to the ceiling to sit on a narrow shelf where they were invisible and waited for visitors to walk past.

Through the tube came a party of sightseers chatting and laughing, until they hit the blacked-out section. Suddenly there was silence, except for some man who complained that the National Park Staff wasn't on the ball with maintenance.

In the stillness, a soft fluttering filled the dank air. (The youngsters were flapping the material on their pant legs.) Women moved closer to men and children began to clutch at adult hands and whimper as the sound increased. Then one women thrashed her arms around her head. She screamed and uttered the explosive cry "BATS!"

Fear. Panic. Soft balls of something were flying around. Women's hair seemed to be their specific target. The flapping echoed everywhere. People ran. Some crouched on the walkway. Screams. Yells. Kids crying.

One man tore off his shirt as a flying object went down inside his collar. Finally a stampede erupted and the tube was filled with flying arms, legs and wild eyes seeking the daylight.

Hawai'i does have a native bat that lives in trees and is so scarce these days that it is on the endangered list and is rarely mentioned in books on island fauna. But no one could convince those visitors to the Thurston Lava Tube —that the cave wasn't loaded with "bats!"

Today the lights are imbedded into the stone so no-one can tamper with them. The bats all went to college, got jobs, married, gave me 10 grandchildren and live spread out across the U.S. Mainland. And that leaves just me walking through an empty cave, not inspired by the works of Pele, but listening for the flutter of wings.

⑥ Devastation Trail

Next stop along the Crater Rim Road is a forest devastated by ash fallout from the 1,900 foot-high magma fountain that half-filled Kīlauea Iki Crater in 1959. The hot ash covered an area downwind with the deepest cover. Here the National Park built a boardwalk so you can enjoy the sight as nature slowly rebuilds herself. Statuesque 'ōhi'a tree skeletons rapidly rot away in lonely dignity here. Shoots of ferns, 'ōhelo berries, and other ground cover, such as the pesky blackberry, are gaining a foothold. Excellent examples of tree molds mark where tree trunks burned out, leaving their shape as a hole in the ground.

The walk is about 3,000 feet long and connects two parking lots. The best deal is for someone to drive the car to the next parking lot at the junction of the Chain of Craters Road and let passengers walk just one way.

There is more to discover near the second parking lot, and none of it is along the boardwalk. Here you will notice the outline of the old highway that has now become a hiking trail. Walk it for only a few hundred yards to arrive at a special place.

Yellow moa fern (psiloturn sp.) at left, was used for tea and talcum powder by Hawaiians. Found growing in the wild in Hawai'i, it is descended from plants that grew on earth 350 million years ago. At right, hot ash from the 1959 eruption fell on nearby 'ōhi'a tree forests, leaving tree skeletons like those along aptly named Devastation Trail.

7 The Perfect View

If you had to choose the one perfect spot to view Kīlauea in every aspect of its volcanic glory, where would that be?

You might guess, "Wherever an eruption might be happening." Maybe so, but it's not a satisfactory answer since eruptions in or around the caldera don't happen often, and those that pop out down its flank are inaccessible by foot or car until Pele sends lava flows down along Puna's shore and into the ocean.

We have to face up to it. Kīlauea is not the most active volcano in the world, but it is *one* of the most active in the world. At the moment it may be the most monitored by volcanology babysitters, but La Fournaise on Reunion Island in the Indian Ocean holds the year-after-year "eruption by the numbers" record for a volcano. You can go there. The island is an overseas Department of France, and is serviced by two 747 flights daily.

Kīlauea's summit area, which visitors come to see, puts on a cold but steaming face much of the time.

So where is Kīlauea's perfect vantage point? For me, it's along Kīlauea Iki's Devastation Trail. One view comes together in all its glory only three or four times a year sometime between January and April.

I'm not talking about the boardwalk trail that was built to see the 'ōhi'a forest devastation wrought by falling ash from Kīlauea Iki's 1960 eruption. The trail I take goes behind that cinder cone.

It starts at the parking lot opposite the beginning of the Chain of Craters road and follows the old Crater Rim Road that was buried with the 'ōhi'a forest. You'll notice a speed limit sign still standing but now it only regulates foot traffic. Flanking both sides of the path are clumps of blackberry bushes. These gradually thin out to a more interesting landscape where dead-white 'ōhi'a stand or are strewn over a plain of mottled brown, black, gray and rusty-colored cinder. Bundles of red

The regal nēnē, or Hawai'i goose (Branta sandwicensis), makes rare appearances near Kīlauea Crater. Hawai'i's most famous endangered species often dines on 'ōhelo berries (next page).

'ōhelo lend color as the berries present clusters of wine red, peach and yellow.

Up the trail, a line of craters pockmarks the landscape. The original land underfoot was probably a gully and the new blanket of cinders is now slowly filtering down. The trail comes to an overlook before switching back down a hill onto the shelf of land separating Kīlauea Iki Crater from Kīlauea Caldera. This ridge is my choice viewing spot.

From here you view most of Kīlauea's Caldera and around three of its rims, practically the entire profile of Mauna Loa and the craggy top of Mauna Kea.

But there's more. The plus is what I call the "flavor of the sight." All around the foreground is evidence of Pele's antics with both a destructive and an artistic touch. It almost looks as if she had planned the scenery deliberately. Naked trees throwing their arms up as if signaling for help; new growth promises a greener world.

Cinders laced with polished drops of Pele's tears gleam gloss black like licorice drops; the curvature of hills built by a rain of wind-blown magma.

Destruction now making a comeback leads the eye to a lush gray-green forest, then downward to the depth of Kīlauea's caldera. The volcano-top depression seems huge until I lift my eyes to Mauna Loa's shape. Over the left shoulder of that mass, Mauna Kea may be more than 50 miles away, but it still looms clear and majestic.

The flavorable accents are exciting if they are present all at once, and only once have they been there for me: Six nēnē eating 'ōhelo and not at all timid as they merely glance my way and shake the yellow, red, and blue bands on their legs; wispy steam coming off the cold caldera floor, clouds of it rising from the north rim steam vents as if the bank was experiencing a forest fire; Mauna Loa wearing snow like a Jewish "yarmulka" and Mauna Kea sporting it as a lumpy old fisherman's hat.

Pele's Sacred Snack

We have to pause here to consider a few species of berries that you have seen. Black is beautiful and delicious. Once a year I harvest volcano black berries. My wife makes them into jam and pies.

The common blackberry around here is the pesty variety that shows up uninvited when land is burned off or its natural growth cut back. Two areas of Kīlauea abound with this berry: Devastation Trail landscape and the hillsides beyond Bird Park on the Mauna Loa Strip Road.

The other black is one type of 'ōhelo. It's not exactly black but a rich midnight blue. And you thought 'ōhelo berries only came in reds, pinks and yellows! No, there is a blue-black and it's mainly found at the higher altitudes. My patch is at the 6,000-foot level of Mauna Loa.

'Ōhelo are Pele's sacred snack. I wonder if the black variety is super-special to the lady? It appears that way to the Hawaii goose and pheasants who will ignore a bush burdened with red juice-filled marbles and feast on the blacks as if there were no tomorrow. Because of this, I only pick high on the bush, leaving plenty within reach of my short-legged friends.

Oh, wow! I almost forgot. Before touching an 'ōhelo you have to perform a rite. It is imperative to acknowledge that these belong to Pele. The procedure, as explained by the Hawaiians to the first white men who ate 'ōhelo, is simple. Pluck a cluster of berries, divide the bunch, face toward the greatest concentration of volcanic activity (whether it's just steam or an actual eruption) and say, "Pele, here are your 'ōhelo. I offer some to you, and some I also eat." Then toss the goddess her share and you're in business.

There aren't enough 'ōhelo around for commercial picking (and you can't make a business of it anyway because the Park Service won't let you) so only a few Volcano residents excel in using the berries. Queen among these was Mrs. Chester Wentworth. Some 40 years ago she invited me to Thanksgiving dinner and served an 'ōhelo berry sauce in place of the traditional cranberry. I made a pig of myself.

Since this booklet is giving up all my secrets on Kīlauea, I'll share the inside dope on making good 'ōhelo pies, jams, and toppings for ice cream or cheesecake. There are two kinds of 'ōhelo, the more common low bush, growing on the ash, and the high bush found in forests. The high bush variety are slightly bitter. Include some of these when cooking to enhance the flavor. Also a little minced apple, a squeeze of lemon juice and a table spoon of crushed pineapple won't hurt.

And now to the blackberries. Whereas 'ōhelo picking is usually a solo operation, blackberries are a community affair. The chances are you'll meet your neighbors sneaking around the stickery clumps plotting to get the fattest berries before you do. Again, the rule is to leave berries next to the ground for the nēnē, a rule little kids fail to follow.

8 Keanakāko'i Crater

Let's proceed along the Crater Rim Road. A pit crater on your left—Keanakāko'i–and the higher ridge wave of ash beyond, together with the foreground of Kīlauea Caldera on your right that the visitor's platform overlooks, provide an important lesson in geology spanning 2,000 years.

Most books on touring Kīlauea gloss over this spot by simply saying that there was a quarry of basaltic glass near the bottom of Keanakāko'i which was mined by the ancient Hawaiians to make spear points, adzes and other stone tools. The quarry was buried by a lava flow in 1877.

For starters this area is the beginning of all the jumpy nerves and arteries flowing with molten rock that form the beginning of the gigantic weakness in this volcano that runs completely down its eastern side and 70 miles out into the ocean. In short, you are standing smack atop the beginning of Hawaii's most active volcanic rift. If the geological clock didn't run so slowly where minutes were centuries, I would say we had better get out of here!

The evidences of a fresh eruption around you occurred in 1974. You can follow its black progress from the caldera floor as the lava exploded from the rift up over the road, through Keanakāko'i Crater and beyond. Scientists look at Keanakāko'i and see much more.

Beginning sometime before the birth of Christ they see another hole in the ground nearby called Powers Caldera. After this one filled with lava they figure a tremendous collapse happened and this begat the beginning of Kīlauea Caldera. In 1790 Keanakāko'i did a minor imitation of Mt. St. Helens and the ash blowout built all these ash ridges, covered the Ka'u desert killing one-third of a Hawaiian army, thereby changing the course of Hawaiian history and blanketing much of the land toward Hilo, including Volcano village. I mean, what are we doing here? I can hear the clock ticking. Let's get on toward Halema'uma'u on the caldera floor.

The Crater Rim Road now dips onto the floor of the caldera. Let's stop at the pull-off overlooking this area because I want you to return here this evening, weather permitting, which means if the heavens are clear.

The trials I went through to find the perfect sunset seat in and around Kīlauea!

At first glance this looks like a bunch of tourists sitting atop a mound of pāhoehoe lava. Look again. This is a rare photo of America's first seven astronauts and their geology instructors from the United States Geological Survey. The astronauts trained on Kīlauea to help them understand geological features when they landed on the moon.

Pictured are:
Donald K. Slayton,
Alan B. Shepard,
Walter M. Schirra,
John H. Glenn Jr,
Gordon L. Cooper Jr.,
M. Scott Carpenter,
and Virgil L. Grissom.

45

⑨ The Sunset Spot

This setting overlooking Halemaʻumaʻu is for the senses. The eyes had first priority. They wanted to watch the sun exit our atmosphere and note the chameleon landscape as it was dressed by the fading light.

My body had to feel the evening coolness wash in and drench the volcano. The senses had to be aware of a new dimension to Kīlauea's caldera. One that whispered "mystery."

The nose needed the whiff of scorched rock, stir-fried air, unearthly gasses coming up in unmannered belches from upset geological stomachs.

The ears were low on the scale of importance. They could be easily satisfied, as they demand total silence like one seeks while alone in the Grand Canyon or Alaska's tundra. Only under such circumstances could they claim to hear the last rays of the sun hitting and scraping the cold blue lava, and evaporating into a misty cloud the Hawaiians call *ohu*.

The weather didn't cooperate every evening while I searched for the perfect sunset spot. Often I couldn't tell if Mauna Loa still existed as an impressive backdrop, let alone that there was a sun to set. For a year I walked all the compass points of the caldera and its rim, searching.

One day I found my spot. What galled me was that it was near a tourist vantage pull-off right beside the Crater Rim Road! I envisioned it would be a secret hideaway that took some trouble to get to, but no, it was about 200 feet off the road. The September 1982 lava flow to the southeast of Halemaʻumaʻu nearly blanketed the area, but left my spot uncovered.

I don't have an exclusive on this spot. You're invited anytime. Hopefully, you'll be alone; the sunset performance somehow demands that. It's a one-on-one happening—just you and Pele.

When you first arrive, hopefully about 20 minutes before the sun connects with Mauna Loa, Pele will attempt to intimidate you. Some people, being alone, are frightened by the vastness of it all. A person is so small out here!

The seat is as comfortable as a smooth rock can be. Behind and a little uphill of you is that darned road, but then there's no traffic about dusk. Beside and before you is the caldera with its sometimes-hot spot crater Halemaʻumaʻu, the darkening line of Kīlauea's western rim, and beyond, the loaf that is massive Mauna Loa.

First off, you notice you aren't being mesmerized by any ONE scene like watching the sun set on the Kona horizon. Here, scenery colors will change three times before dark. Your eyes and senses will be bouncing around the foreground and sweeping the distances. But always they will return to the sun and finally zero in on that gem as if it were a valuable ring going down the sink drain.

I chose this one spot near Kīlauea volcanic action becauseI can feel Pele breathing. From the odor you know she should take a bath tonight, and lately she has been doing just that off the Chain of Craters road.

The wind at my back comes off the beginnings of the Ka'u Desert bearing no smells. It will stiffen, become cool, then refrigerator-fresh after the sun disappears, so better to wear a sweater.

Immediately across the foreground are two rips of past eruption rifts (1974 and 1975) that are marked with spatter cones wearing cloaks of sulphur ranging from yellow to white. They puff silently, their breath growing denser as the cool evening comes on. By the end of the sunset show this foreground will all be inked in and the cones silhouetted.

The open fish mouth of Halema'uma'u shows its hot lips. In the end this gigantic hole will blend into the caldera floor but its breath will remain.

Beyond, the darkening rim of Kīlauea increasingly emphasizes its man-made flaw: the Volcano Observatory atop the high ground. As darkness dims natural light, electric lights will brighten windows that are alien to this scene.

Mauna Loa at first is tan with a green fringe on her skirt. Gradually it blushes pink, then deepens to a lavender, and finally it too becomes a silhouetted logo for the most massive single mountain on earth.

And now the sun. Hurts one's eyes to be rude and stare at it at first, but it seems to lose power in fits and starts as it nears the summit of Mauna Loa. It acts like my car's ignition system: goes great, then stalls, then powers on again. Well, the sun does this as it approaches the volcano. The final stall happens upon reaching the summit. There it just sits, and sits.

"In fact, It's stuck!" I tell the 'ōhelo bush next to me. See! The 14,000-foot mountain has some kind of magnetic hold on the sun. I've noticed it so often that I have come to believe it. And that's okay by me because the sun sitting on the very rim is a diamond with a yellow fire burning within its crystal depths. Even when the sun's ignition gets working again and it slides out of sight, the sparkle remains awhile.

A different scene takes over when there are trails of wispy clouds high up. The clouds will deepen in color from pink to red. In the end the sky turns lavender deepening to purple.

These colors in turn will be reflected in a muted way off cold caldera lava, and to your right, off the amber blister of Kīlauea Iki's cinder dome.

You can sit there longer if you wish, but it's only going to get colder and lonelier. Pele's gone deep into her fissures. She'll return when a full moon is out around midnight to skate on the pāhoehoe.

⑩ Halemauʻmau

Of all the interpreters of Hawaiiana over the last 150 years, only three have recognized there were two types of Halemaʻumaʻu. The only one known today is the "hot-spot" on Kīlauea's floor. The others, and there were several, are in the forest on the volcano's east rift zone.

Halemauʻmau Means "House That Endures"

The Hawaiian word *halemaumau* has two different meanings. It is pronounced differently for each. When pronounced Halemauʻmau it means "house that endures," which is the historic name established for this crater on Kīlauea's floor by a Polish geologist in 1838. Back then this Halemauʻmau was in constant eruption.

Halemaʻumaʻu Means "House of Ferns"

The other pronunciation, halemaʻumaʻu, means "House of Ferns," and this ancient name is applied to fern-choked pit craters along the Chain of Craters rift. There is a particular fern named *amaʻu* whose new fronds unfold a brilliant orange-red in the spring, resembling a fountain of molten rock.

There was a third Halemaumau, but do not confuse it with a crater. It was the name of a legendary *kahuna* (Hawaiian priest or sorcerer) from Kauaʻi who arrived at Kīlauea with a contract to kill Pele. He lived in a hut made of the fronds of the amaʻu fern near where the Volcano House hotel now stands. Well into the first

49

third of the nineteenth century any temporary shelter of native vegetation built on Kīlauea's rim was called a halemaʻumaʻu or "house made of ferns."

Halemauʻmau, the house of enduring fire, is not so eternal these days. But it is still impressive, especially at dawn when I see its breath in the cool morning air.

Sunrise across the caldera floor is the only time the face of Halemauʻmau has on makeup. Since I share the view of the Hawaiians that this crater is the present home of Madame Pele, I think of this area as dynamically feminine. But not too friendly, I might add. There's a keen edge on my sensibilities when I explore this property. I've been around when Pele's basement was far deeper, when it was floored in molten rock, issued vile gasses, and when the occupant came to the door and demanded "Who's there?" Her breath scorched wooden viewing platforms and blistered paint on automobiles.

The orange-red amaʻu fern.

Perhaps this is the time to tell you that the only deaths by volcanic action in modern times occurred right here. A man photographing the steam eruption from Halemauʻmau in 1924 had his legs crushed by boulders flying out of the crater. He died on the operating table at Hilo Hospital hours later. In 1987 a lady visitor died instantly right where you're standing when she succumbed to sulphur fumes.

But now on a cool morning, this unpredictable goddess is only breathing hard in an uneasy sleep. She went to bed without removing her makeup after partying the night before. Her lips are brushed faintly red. Yellow sulphur is rubbed on both cheek bones. Her hair still jet black but with ribbons of silver and burnt orange. Around Pele's bed are smoking pots of incense fuming in dawn's light.

For years now I've tried to capture this scene on film. The time span for success is so short —maybe 15 minutes between dawn, when there's light enough to expose film to the moment when the glaring sun comes over Kīlauea's rim. The secret is to shoot into the light to get the effect of sulphur, steam, sunlight's first rays and silhouettes of primeval volcano shapes.

Excellent examples of halemaʻumaʻu, home of the ferns, are two pit craters on either side of the Thurston Lava Tube. Of course the popular tube's basin is a halemaʻumaʻu too, but it is contaminated by people who talk, laugh, chew gum, and jingle car keys and small change in their pockets.

⑪ Southwest Rift Zone

Rising back to the rim of Kīlauea after Halemau'mau, one can see evidences of both the beginning of the Southwest Rift Zone of this volcano and the Ka'u Desert on your left. This 15-mile-long area is sparsely visited even by backpackers.

The U.S. Geological Service Observatory on the highest portion of the rim called Uwēkahuna Bluff is well worth a half hour stop. A new section of the observatory was completed in 1987.

⑫ Jaggar Museum

The older building has been turned into a minor museum of volcanology named after Thomas Jagger, the first resident scientist to study Kīlauea beginning in 1912. Restrooms and drinking water are available.

While progress adds much, it also takes away something that was cherished. The old building crammed with equipment, charts, dusty bookshelves, and study tables littered with ongoing projects made the scientists highly visible to their public. It was a comfortable feeling to see that someone was actually baby-sitting the volcano.

Then too, the building became a focal point for

Volcano residents whenever an eruption occurred. The word of Pele awakening and doing calisthenics would flash through the village. Friends phoned friends and within minutes of an eruption, cars would stream out of driveways and converge on the observatory. The time of day or night didn't matter. Even in the wee hours of a dark and cold pre-morning we would stand around in pajamas under coats, clutching a thermos of coffee, waiting for a scientist who would periodically stick his head out the observatory door with the latest scoop.

Kīlauea Military Camp

Those buildings on your left make up a rest and recreation complex for active duty and retired military personnel. All Volcano residents benefit from their fire and ambulance service, and many of their recreation programs. Over the past few years, portions of the complex have increasingly come under public contract, such as the cafeteria, bowling alley, and gas pumps.

Steaming Bluffs and Sulphur Bank

For those of you not taking the early morning stroll through this area, you can now motor past these sights. Please keep the kids under control around the steaming vents and don't harvest sulphur crystals.

To visit the upland area of Kīlauea along the slope of Mauna Loa return to Route 11 and turn south. This part of the park does not demand that you pay to enter.

Mauna Loa Strip Road

Most visitors skip the tree molds at the beginning of this drive. Holes in the ground showing the size of prehistoric koa trees aren't that impressive, not when you can see one of these giant trees at the next stop.

Bird Park and Kīpuka-Puaulu

Bird Park is one of the more delightful picnic parks at Kīlauea. Next door is Kīpuka-puaulu, a 100-acre chunk of land that was left in its forested state when a lava flow in 1560 came through one side of this plot and another flow in 1600 hemmed in the other side. Descendants of some of what is still growing in the kīpuka can be traced back to about 200BC. This should make the one-mile loop walk through this forest botanically exciting, especially when the National Park claims "it has one of the richest concentrations of native plants in Hawai'i," not counting the fact that there are a few of our native birds peeping in the trees. But alas, guidance material necessary for you to identify, understand and put into context what you will see on the walk has always been poor. Have a good walk anyway.

⓭ Mauna Loa

The narrow, winding road beyond Bird Park goes for 11 miles up the side of Mauna Loa to a parking area for the start of the hiking trail to the summit of this volcano. It's mostly wooded with new and aged clumps of koa trees. The only vista is at the 6,000 foot level where on clear days, the broad shoulder of Kīlauea can be viewed most of its length down the east rift. The road ends at a turnaround where the 19-mile trail to the summit of Mauna Loa begins. This is a two to three day hike with overnight cabins at the 10,000 foot level and near the summit. Winter is a bad time to take the hike.

Steaming Bluffs

⑭ The Seasons at Kīlauea

Visitors often remark that Hawai'i basically has a one-season climate that is monotonous. Residents of Hawai'i who have lived in the tropics long enough realize that we do have two definite seasons, summer and winter. The botanical aspects of this drive show these climatic changes, as does Kīlauea in general. Let me introduce you to our seasons on the summit of Kīlauea.

Summer at Kīlauea

Summer at Kīlauea Volcano started at 1pm today, July 9. I know, because the birds are suddenly thirsty.

Birds are the only reliable harbingers of the beginning of summer at my home atop Pele's world. Flowers blooming in their season, or a lesser rainfall, or temperatures staying above 70 degrees are not true indicators of a change in our two-season climate.

You learn to watch the native forest birds like the *'amakihi*, *'i'iwi* and *'apapane*, and the Japanese tourist who has found a home here, the *mejiro*. They all have a varied diet of insects and nectar; and it's their nectar consumption that says summer has arrived.

I have a saloon outside my bedroom window. It's a 40-foot 'ōhi'a tree with its flowering lehua. This flower is the watering hole of life for Hawai'i's forest birds of the Hawaiian honeycreeper subfamily.

You'll probably get the idea that I have nothing better to do than watch birds stick their beaks into lehua blossoms. It's fascinating, really, and becomes exciting when summer arrives and the birds have a happy hour every hour.

I've been wondering, how quickly a lehua replenishes its nectar load. The poor tree's systems must be going into overtime, because one bird after another visits the same flower time after time. The door to the bar never stops swinging as they come and go, and those leaving seem be satisfied.

I pored through all the books available to me on nectar versus birds but found no answer, so I phoned the State's Wildlife Service.

"Hi. How often does a flower reload its nectar after a bird has sucked it dry?" I asked.

"Whoo boy! You would come up with a tough one, wouldn't you?" said the voice. "We're more into wildlife here than plants or flowers."

The bewitching Hawaiian ʻapapane perched on a branch of Pele's own tree, the ʻōhiʻa lehua.

First, you have to visualize what a liter of liquid is. Then try to guess what makes up one ten-thousandth of a liter, which is called a "micro-liter." I suppose the tear film left on your eyeball after you blink is a micro–liter.

Well, a lehua blossom serves up .75 of a micro-liter of nectar per hour during the day and .42 per hour during the night. I can't begin to get a grip on that amount beyond knowing that a honeybee gets loaded up in two hours. For a bird, 100 times larger, it must take all day!

Anyway, someone has figured out that an ʻōhiʻa tree the size of mine, with the normal load of several hundred blossoms, is a daily restaurant for eleven birds. That is, until it rains. Then the nectar is watered down to a cheap drink and the bird has to guzzle twice as much to get its calories.

How did we get on this subject anyway? Oh yes, summer started this afternoon. I knew it for sure, because my fast food tree that has seating for eleven, is now handling 36 thirsty birds.

Winter in Hawaii?

Yes. It arrives at my place at Kīlauea one day a year and stays around for seven hours and 19 minutes. The exact day is unpredictable. It's usually sometime in January or February. This year it was January 16.

There is no foretelling this winter spell. The

"Well, how many times a day does a forest bird sip nectar?"

"Whoo boy! I don't think any researcher has been able to keep their eyes on the same bird all day."

"Well then, does the same bird go back to the same blossom and drink twice? Or does he frequent a different pub each time?"

"Whoo boy!"

"Who am I speaking to?"

"I'm just the janitor around here. Everyone's out in the forest studying animals and birds." I said I hoped he would have a nice day and hung up.

The answers to my questions apparently aren't for general human consumption. I finally discovered them in a study paper intended for research eyes only.

cat doesn't come to bed with me. The spikes of my cymbidium orchids continue to wave their pink, coral, and red blooms. I'm thinking of mowing the grass but I don't because of threatening rain. Of course, the weatherman doesn't give a clue as to any winter lurking.

My tropical winter starts in the dark hours when I roll over in bed and glance at the red glow of the clock. Two o'clock. What woke me? Ah, my cold nose. That's my indicator. I know what to expect, come dawn.

I have never known my winter morning to be anything other than so clear it hurts to take too deep a breath. My living room is chilled and I build a fire in the 1868 Franklin fireplace. It's either that or my wife stays in bed and breakfast is late.

I open the front door and the cold surges in like an ocean wave through a reef blowhole. The temperature reading on the porch is 36 degrees. I imagine I hear crying out in the yard. Which of my plants can't take this winter morning? Perhaps it's the bird of paradise. No,

it's real birds, mewing in the 'ōhi'a trees as they flit from branch to twig with a bounce that shakes the cold dew from scarlet lehua blossoms. No sense in sticking their beaks into ice water before feeding.

The dog is gazing into his water dish. Probably he's amazed at the dog looking back from the plastic-thin sheet of ice on the surface.

My ocean of lawn has a faint spray of silver on it. Each

blade of grass is trimmed in frost. Except, of course where the dog has explored, leaving a wet, green trail.

There are some happy faces in my cold yard. Pink eyes of camellias wink at me from among glossy green leaves. Blushing rhubarb stand tall holding umbrellas of leaves.

Chalices of calla lilies are turning their frost-dusting into droplets of water that dribble into their white throats. Several hundred pink Cymbidium orchids on a hillside are waving and cheering because this is their type of weather.

Anthuriums are among the sad ones. What few heart-shaped flowers there are during this time of year are now rimmed with an almost transparent purple blister of frost burn. Ginger leaves limp earthward and California poppy blossoms look so tight-fisted I doubt if they will open today. No bees are airborne. Perhaps their queen has grounded them until the sun warms up the world.

I go to the far corner of my property to view the traditional winter I know is on display. There I see the summits of Mauna Kea and Mauna Loa. Kea's white hat is lumpy, covering the summit cinder cones, and floppy as it veils down to the 10,000–foot level. Loa wears a quilt of snow on its smoothly–rounded dome. The cold breath from neighboring Mauna Loa has invaded our 4,000-foot Volcano Village at Kīlauea.

The sun finally decides we have had enough winter. First it shafts through the surrounding ʻōhiʻa forest, bounces off my stiff house, then drenches the shivering yard. Our corrugated iron roof snaps and crackles as it loses its cold. The bamboo steams silently as if a cloud had been trapped within.

I return outside after breakfast to find the temperature 50 and rising.

"Well, that's it for another winter," says my wife.

A Room With A View . . . Japanese maple in full bloom in Gordon's garden as seen from the breakfast room in the historic house at My Island Inn.

Daring Pele.
Don't try this at home!

Austin and Anneka, two
young visitors from St.
George, Utah, pretend to
make a dash for it in front
of an oncoming flow that
chose the easiest route to
the sea, a manmade road.

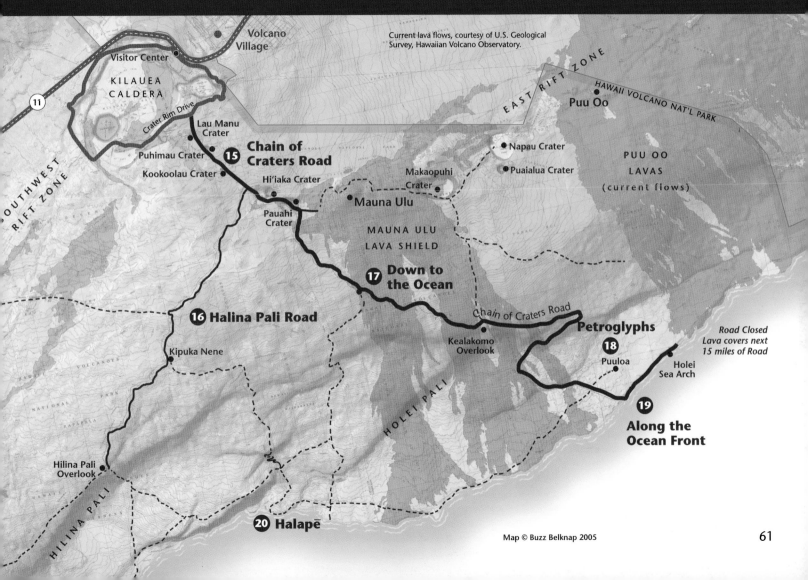

Current lava flows, courtesy of U.S. Geological Survey, Hawaiian Volcano Observatory.

Volcano Village

Visitor Center

KILAUEA CALDERA

11

Crater Rim Drive

SOUTHWEST RIFT ZONE

Lau Manu Crater

Puhimau Crater

Kookoolau Crater

15 **Chain of Craters Road**

Hi'iaka Crater

Pauahi Crater

Mauna Ulu

MAUNA ULU LAVA SHIELD

Makaopuhi Crater

EAST RIFT ZONE

HAWAII VOLCANO NAT'L PARK

Puu Oo

Napau Crater

Puaialua Crater

PUU OO LAVAS (current flows)

17 **Down to the Ocean**

16 **Halina Pali Road**

Kipuka Nene

Chain of Craters Road

Kealakomo Overlook

HOLEI PALI

Petroglyphs

18

Puuloa

Road Closed Lava covers next 15 miles of Road

Holei Sea Arch

19

Along the Ocean Front

Hilina Pali Overlook

HILINA PALI

20 **Halapē**

Map © Buzz Belknap 2005

Tree molds, eerie, otherworldly remnants of a vast 'ōhi'a forest, emerge as lava slakes slowly seaward during the Mauna Ulu eruption of the 1960s.

15 Chain of Craters

After experiencing Crater Rim Road and all its goodies, the short-time visitor asks himself: "Do I have time for the Chain of Craters Road? Is it worthwhile?"

You bet. This road and all of the lower Puna district which comprises the flank and lowlands of Kīlauea, are just as important in getting to understand this volcano, and more diverse in geological spectaculars than some of the sights at the summit.

This then brings up the situation that you may have misjudged the Big Island of Hawaii and have not budgeted enough time here. Happy solution: cancel time on one or more of the other islands and extend your visit here. Go ahead. That's why they invented the telephone. An airline reservation is easy to change. Didn't you know that the Big Island of Hawai'i is the only island worth visiting?

The first four miles of the Chain of Craters drive is pocked by impressive holes in the earth called pit craters. Several were probably created when Kīlauea's summit collapsed, others much earlier between 1400 and 1700AD when so much magma came out of the plumbing system under this rift that the ground sank here and there to fill the void. Since 1955, activity on this rift has resumed and lava has welled up across a few of the pit craters' floors or exploded from areas in between.

The sight of these pit craters, and the knowledge of how they were formed should be cause for serious reflection. Since 1969 when Mauna Ulu was formed, and now with the ongoing action down at Pu'u Ō'ō, so much material is again coming out of the earth that scientists wonder if Pele will stamp out another pit crater or two, or even recollapse Kīlauea's caldera floor.

16 Hilina Pali Road

But let's keep some happy thoughts as we drive past Lua Manu and Puhimau Craters. Off to the right will be Hilina Pali Road, a narrow winding road that goes eight miles to a cliff overlooking the southern flank of Kīlauea. Very few visitors make this jaunt as it really doesn't offer that much in the way of new scenery. To someone interested in geology it is important as the finest examples of faulting in the Hawaiian Islands are found beside the first three miles of the road. A major clue to the future of Kīlauea is within these faults. Evidence shows that the entire south-east flank of this volcano is beginning to slip into the ocean. Someday there is expected to be an avalanche where the resulting tsunami

to Kalapana along these pit craters. In 1965 the National Park enlarged the road and ran it all the way to the seacoast. Eruptions from Mauna Ulu cut the highway in 1969 just beyond the Mauna Ulu parking lot. In 1979 the road was rebuilt along a different route to link up with the ocean-side section.

When M auna Ulu erupted in the late 1960s into the 1970s, it was the most sustained outpouring of lava in Kīlauea's recorded history to that time. You'll soon be riding through d went over a cliff all the way into the ocean.

Today a 3.2 mile (round trip) trail leads from just beyond the parking lot to the top of Puʻu Huluhulu, an ancient cone behind Mauna Ulu. The walk is worth your time. One of the most rewarding panoramas on Kīlauea is seen from the top of Puʻu Huluhulu where a bronze marker identifies the landmarks on a 360-degree sweep.

Be alert during this walk and when driving over the next three miles. The goddess of volcanoes has something to show you.

might devastate the Hawaiian Islands, and possibly affect all land around the Pacific rim. Along the cliff face at the end of the road can be found the oldest lava on Kīlauea, dating from 20,000 to 100,000 years ago.

Backpackers interested in hiking to the southern coast of Kīlauea go as far as Kīpuka Nēnē, which offers a trail head. Back on the Chain of Craters Road we finally come to rest in the parking lot of Mauna Ulu.

The upper section of the Chain of Crater's Road generally follows the ancient trail from Kīlauea's summit

⑰ Down to the Ocean

Now down toward the coast. To those who dig geology, this drive is a must. The landscape is a frozen ocean with waves, currents, wind-tossed water and all. The secret to photographing this spectacular scene is to go through the area before 9am or after 3pm, when the sun shines with a slant upon the lava's surface. The scene changes every way you turn.

A good place to stretch your legs and use up some film is at the overlook shelter at Kealakomo. Here you view the Pacific from atop the 1,200-foot-high Hōlei Pali where Mauna Ulu's lava flows waterfalled over and down across the coastal plain to extend the island another quarter-mile toward California.

While you're gazing out to sea, you might like to know that someday a new Hawaiian island will appear out of the ocean 18 miles offshore. Alvin, the research submersible that photographed the *Titanic*, has already explored this underwater volcano. *Lo'ihi* has only 3,000 feet to go before it breaks the surface, possibly within the next 20,000 years. If you think that is a long way to go, consider this: That submarine volcano is already 10,300 feet high. It has a caldera roughly the size of Kīlauea's with two craters resembling Halema'uma'u and Kīlauea Iki inside of it!

Down at the bottom of the pali where the road curves left toward the ocean is a wide pullout. Look to the uphill side of the road to see graphic examples of both a'a and pāhoehoe lavas side by side. On the downhill side, the flow covered the old highway and pieces of the macadam with its yellow center line are still showing.

From the overlook shelter at Kealakomo it's easy to see how volcanic activity has expanded Hawai'i in this century. Just beyond the horizon here, at a mere 3,000 feet beneath the sea, Pele is building Lo'ihi, the next island in the chain.

⑱ The Petroglyphs of Pu'uloa

As the road levels out a mile further on, a pull-off and a sign show the way to the Pu'uloa petroglyph field. The round trip walk is 1.8 miles. Today we'll visit the petroglyph field at Pu'uloa. What we will see in Hawaiian rock art is radically different from everyone else's views. There are about 20,000 petroglyphs here.

Pu'uloa means "Hill of Long Life" in Hawaiian. It isn't a hill but more like a blister of pāhoehoe out in a vast lava flow that Pele laid over the landscape in the fifteenth century.

The trail to the petroglyphs from the Chain of Crater's road is about a mile of easy, level walking. This path has a unique Hawaiian beauty as it meanders around and over waves of lava. It was worn into existence by human feet, some barefoot and some shod in woven leaves that left the rock contoured and varnished like much-trodden marble steps in ancient Greece.

In former times the trail was part of the "round the island" thoroughfare. Why that particular blister of lava was singled out for Hawai'i's largest field of art in rock is not clear. Perhaps because there was a major trail with a junction nearby that led to Kīlauea's summit. Or because an important religious temple was nearby.

At any rate, this bubble of lava is pock-marked by thousands of holes the size and depth of a teacup.

Anthropologists tell us the holes were chipped by the parent of a newborn child who brought the umbilical cord here to deposit in the hole and hoped the gods would grant the child a fruitful long life. They call them *piko* holes. Piko is your belly button, or umbilicus.

But I'm after more meaningful stuff, and I find it around the perimeter of the hill. Here are impressions of human forms, sea life, items of importance in everyday existence, and what looks like just plain "doodles" made as a traveler rested along the trail. These are scattered over many acres. The preferred slates for this art cut into rock are islands of flat, sun-mirrored slabs of pāhoehoe. The favorite pastime of those who enjoy this expression of former humanistic art is to guess at what the pictures mean. A fish, crab, or mat sail of a canoe seems obvious, but the childish stick figures of a man or woman offer complications. Did this represent a particular person? Is some deed or event by that person indirectly implied here? The feeling that a greater story surrounds this particular chipping in the rock invades my senses.

My wife Joann is obsessed by any petroglyph or pictograph left by any society, and she once asked a leading anthroplogist, "What do they mean?"

"Were you there when it was originally cut into the rock?"
"No."

"Well, neither was I. Only the artist was absolutely sure why he did it and knew its true story or meaning. Forever and ever the rest of us will only be guessing. If you are educated as to the life style of these ancients, then your guess is as good as the next person's."

And so I walk around playing my own guessing games. This is where I see these petroglyphs as no current researcher into Hawaiian history sees them.

First off, I do not visualize this rock art as "ancient." That word is misused in Hawaiiana as if everything predating Captain James Cook's Owyhee of 1779 is ancient. To most historians dealing with areas other than Pacific islands, the era before Christ is the ancient parade of mankind, or more specifically, the world of an aboriginal tribe in a given locality. The Polynesians who migrated here in organized families were not the aborigines of Hawai'i.

This reminds me of a humorous anecdote. At a country store in Kalapana (a community that was destroyed by lava in 1990) was a water well with a sign: "Ancient Hawaiian Well–1896." My mother, who was born in 1896, took one look at that sign and went into the store prepared to slam the goofball over the head with her purse who would have the audacity to call her "ancient." The sign was taken down.

What I'm really saying is would you consider the Pilgrims of Plymouth, Massachusetts, ancients? Neither would I.

What I see in petroglyphs at Pu'uloa is not ancient art shrouded in vague mystery. My time frame here (and that of other historians) is 1500 to 1800AD. Now the human being with the spiritual drive that caused him or her to labor out here in the hot sun with hammer, stone, and sharp-edged rock to cut these symbols, comes to life for me.

It is also exciting to search for a petroglyph that can be identified as "art" — a creation born of the spirit and feeling of those times. A thinking man's doodle which passes on a pictorial message that I can relate to. There are precious few of these in Hawaiian rock art. I can only think of ten out of the thousands pecked in lava.

An artist's rendering of the only known petroglyph of a cockfight in the Hawaiian Islands prior to Western rediscovery in the 18th century. It is in a location that remains undisclosed to the public by the National Park Service and archeologists at Hawaii's Bishop Museum.

Petroglyph Art

The prize among them is here at Puʻuloa. I know just where to find it at the far edge of the complex. I'm standing over it now. It's a rendering of a sailing canoe with the overall dimensions of 15 by 25 inches. The watermelon slice of canoe is 2 by 15 inches and the leaf-woven sail is 19 inches tall.

The tilt of the canoe suggests it is sliding down the backside of a wave, its sail filled with the trade wind. There is joy here, a feeling of exuberance, motion, an ocean adventure, and the secure feeling that the sailor or fisherman knows what he's doing and where's he going.

But there's more—a whole world more. It is traditional that tied to the top of the mast is a hank of dried ti leaves that act as a weather vane telling the direction of the winds. But instead of dried leaves at the top of this stone mast there is an . . . . but what, exactly, is it?

Vision of a Hawaiian Michelangelo

Oh, excuse me. I'm making believe it's three in the afternoon about the year 1687 and I see a man coming down the trail. He's about five foot six inches, on the stocky side of powerful in build with skin darkened to near purple-black by the sun. I would say he's about 25, as his youthfulness has fled, but he's not yet 30 as that would be well past middle-age for the average Hawaiian.

He carries something bundled in a green ti leaf. He's barefoot. I'll go sit over there out of the way and watch what he's up to. I'm making believe this canoe petroglyph doesn't exist yet.

The man goes over to the hill pock-marked with piko holes and looks them over. Ah, now I know. He has a newborn's umbilical cord in the leaf.

But somehow he's not satisfied with the norm. This must be a special child. Surely a boy. Perhaps he's had all girls and this is the only male.

The man leaves the hill and wanders around the surrounding flats, stopping now and then to study a petroglyph. Finally he chooses a table of pāhoehoe and goes down to his knees brushing aside flakes of loose lava and wind-blown ash. From inside the waistband of his malo he pulls two shaped rocks. I see that he holds a hammer stone and a sharpened adz head without the handle. He contemplates the lava for a moment, then sets to work.

He must be both a canoe maker and a fisherman because his industry is fast and sure, his strokes positive with what he sees in his mind's eye. His canoe is pictured, then its sail. With the heavier pounding of determination, he chips a *piko* hole atop the mast. Into this he places the umbilical cord and covers it over with the lava chips.

It's now near evening with the sun gone behind the crest of Kīlauea. He stands over his creation in meditation for a long minute and the vibrations from his spirit, hopes and dreams wash over me. I know he is saying to his gods, "Please grant my son the power and length of life to be the greatest fisherman in Hawai'i."

As the man turns and walks away I marvel at the perception, the creativity, the uniqueness of spirit, the artistic urge that, in his day and age and at the level of his society so alone out in the middle of the Pacific, could come to the surface in such a man. The disappearing fisherman is my Hawaiian Michelangelo. Three hundred years later I am enjoying his art.

The most concentrated collection of petroglyphs in Hawai'i is found at Pu'uloa.

Prominent among beds of pāhoehoe used for the history-telling images here is a sacred mound where the umbilical cords (piko) of infants were deposited. Individual piko locations are easily identified by their circular form (see images at lower right).

Fragile petroglyphs at the Pu'uloa site are best viewed from a convenient wooden platform maintained at the perimeter of these extensive fields.

⑲ Along the Ocean Front

The final several miles of Chain of Craters Road borders the ocean. The water will be the bluest you'll ever see. Two pullouts offer views of natural lava bridges.

When this was written in 2004, the highway ended under a fresh lava flow to the ocean. In the past, 15 miles of the road went onward to Kalapana in lower Puna. All this has been covered by rivers upon rivers of lava since 1986. Besides destroying more than 150 homes, a National Park Visitor's Center, and a scenic pool called Queen's Bath, the flows surrounded, and then buried Waha'ula Heiau. This former temple of worship had never been restored even though it was one of the more important in our state. It dated from about 1250 AD when a religion rich with human sacrifices spread over Hawai'i. Bishop Museum chose this one to recreate in the main room of the Honolulu museum.

All tour books on Kīlauea end where the Chain Of Craters road ends. For some reason no one thinks of the lower Puna district, which is the point beyond, as part of Kīlauea Volcano. I mean, don't you consider your lap part of your body? Puna is important in the overall understanding of Kīlauea, and since 1965 it has become all the more so as heavy volcanic activity moves down the east rift into this area.

Unfortunately, while Pele does her thing across highways, you have to gain access to lower Puna by traveling toward Hilo on Route 11, then taking Route 130 down to Kalapana. We'll continue our tour from the black sand beaches of Kalapana in a moment, but first I want to tell you about a swimming landmark that Pele has destroyed in the barrier of lava before us.

Rope lava making its way to the ocean, crosses a highway and blocks it to drivers in the process.

20 Tsunami at Halapē

If the reader is prepared for a hike with overnight camping, then Halapē is an excellent choice. It's nine miles to this ocean-side spot from either the Chain of Craters road, or Kīpuka Nēnē on the Helina Pali road. To a wanderer who loves the outback, this is paradise. But read on and I'll send a chill up your back as to what Pele does to places like this when she is of a mind to do some fancy foot work like she did on Nov. 29, 1975.

Before this time, Halapē offered a white sand beach fronting a coconut tree grove, a shallowish bay that made for good swimming and snorkling. The National Park offered a shelter and water tank.

On this Thanksgiving weekend night Halapē hosted 32 campers of Boy Scout Troop 77 from Hilo, and a fisherman. At 3:36 a.m. Pele came to visit and did a dance that produced a 5.6 earthquake. Campers awoke to the sounds of rocks falling from a cliff back of the beach. Thinking an avalanche was in the making, they fled to the ocean front. At 4:48 .m. Pele apparently got tired of party dancing and did a Tahitiian shimmy, and as a result created the second largest earthquake in historic times.

The south coast of Kilauea dropped 12 feet in places, and moved a short distance into the ocean.

This earth movement generated a tsunami that raced ashore as much as 300 feet inland and 50 feet above sea level. First wave was a five footer, second wave was 26 feet high. Campers at the waterfront were swept inland on the waves into a ditch against a cliff. The fisherman and the Boy Scout leader were killed, and 17 others suffered injuries.

Across the Big Island this earthquake caused $4.1 million in damage, destroying seven homes, two vehicles, and a section of Kilauea's Crater Rim Road fell into the Caldera.

The before and after photos you see of Halape on these pages were taken by my son Tim when he was a light plane flight instructor at Hilo's airport.

Halapē Campground December 10, 1975, left;
Halapē three months earlier, August 21, 1975, right.

Big Island residents watch the first night's eruption beside the town of Kapoho in 1960.

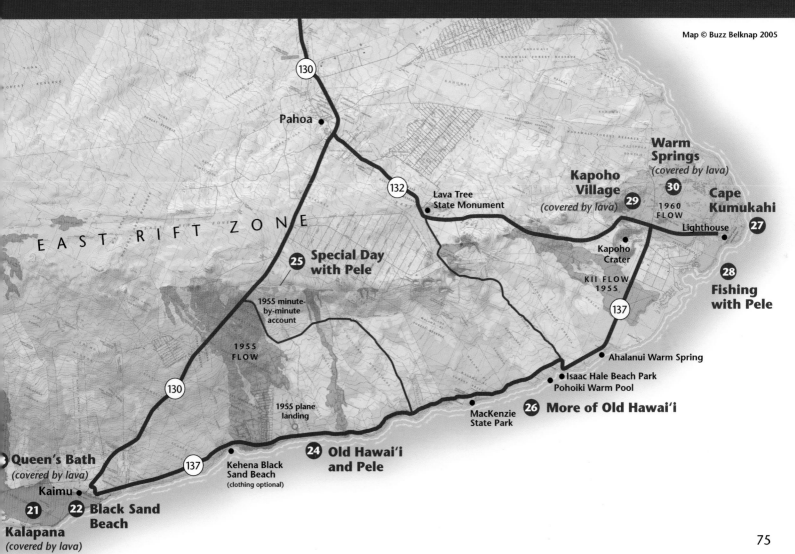

Map © Buzz Belknap 2005

130

Pahoa

132

Lava Tree
State Monument

EAST RIFT ZONE

**Warm
Springs**
(covered by lava)

**Kapoho
Village**
(covered by lava)   29

30

1960
FLOW

**Cape
Kumukahi**

Lighthouse   27

25   **Special Day
with Pele**

Kapoho
Crater

28

**Fishing
with Pele**

1955 minute-
by-minute
account

KII FLOW
1955

1955
FLOW

137

130

Ahalanui Warm Spring

1955 plane
landing

Isaac Hale Beach Park
Pohoiki Warm Pool

26   **More of Old Hawai'i**

MacKenzie
State Park

3 **Queen's Bath**
(covered by lava)

Kaimu

137

Kehena Black
Sand Beach
(clothing optional)

24   **Old Hawai'i
and Pele**

21   22   **Black Sand
Beach**

Kalapana
(covered by lava)

75

Kalapana's Kaimu Black Sand Beach as it looked 1943.

21 Kalapana

From Volcano, travel 19 miles on route 11 toward Hilo to Kea'au, and turn right onto route 130 headed toward Pāhoa and Kalapana. Houses will become more numerous as you continue toward Pāhoa. This area is known as Hawaiian Paradise Park, Orchid Land Estates, 'Āinaloa and Hawaiian Beaches subdivisions. Formerly it was all ranch land, but beginning in 1950 it was subdivided and offered as vacation lots to visitors. Twenty years later many of these buyers began to retire and build homes here. Now young families are moving out of high-priced Hilo and building homes, making this the bedroom community of Hilo.

Approaching Pāhoa there will be acres upon acres on both sides of the highway covered with black netting. These are anthurium nurseries. Go have a look.

Pāhoa started out life in the late 1800s as a lumber town making koa and 'ōhi'a railroad ties, not only for our local rail lines, but also for the Sante Fe Railroad. The cleared land was planted in sugar, so Pāhoa became a sugar town. Today it is the anthurium and papaya capital of the islands.

After Pāhoa, Route 130 crosses Kīlauea's east rift and ends at Kalapana where again you are dead-ended against Pele's paving of the landscape. This area is populated, but homes are being destroyed and a village consisting of two churches, a store, a county park and historical sites was recently covered with lava.

22 Black Sand Beach

The big deal in this area are the black sand beaches. I can still play my game where Kaimu Black Sand Beach used to be, but I can't walk where I used to walk. The beach has disappeared under lava. A few thousand feet out to sea a new black sand beach is being created by Pele.

I once talked to a Hawaiian fisherman who had enjoyed Kaimu's black sands around the turn of this century.

"The beach was out there," he said waving an arm to indicate somewhere out in the middle of the bay.

"You're kidding," I said.

"No, no. The sand was 300 yards further out there," he insisted. "I know. When I beached my canoe it was a long walk up the sand to these coconut trees where my brother lived. We had to carry the fish and nets all that way."

"What happened to all that sand?" I asked.

"Gone out to sea. Pele keeps sinking the island and the sea takes the beach away."

Later I saw a photograph of Kaimu taken about 1912 and my Hawaiian friend was right. In 1946 I photographed the beach. It was much further inland but still a wide crescent of jet black sand stretching from end to end of the small bay. In 1986 I took another picture from the same vantage point. The beach had receded another 50 yards or more. The coconut trees of 40 years ago were gone, and a new crop was now being threatened. Only a tiny strip of sand remained at the southern end of the bay. All the rest was rocky headlands.

The glory of that beach was its utter black and blinding white boundary that changes every minute. This is where blue ocean water turned into a green wave that exploded into foam and a sheet of froth swept up the black beach to where its energy ended in a snowy edge on the blackness. The suds quickly dissolved leaving the sand to spangle in the sun until the next whiteout happened and left a boundary in a different place.

This is where I played my game. I took a stick and drew a line, saying, "You only go to here, ocean."

But I'm no Moses and cannot command the sea. The wave's next whitewash erased my line and I swear the surf laughed. Sometimes it didn't even reach my mark and I laughed back. Then the next wave came faster and reached higher than I expected. It swirled around my ankles and filled my shoes with a chortle. I did not laugh.

I sat on a rock to dry my shoes. A tour bus rumbled down the road and stopped with a hiss. The door opened and out poured a stream of visitors who lined up at the beach edge to point cameras.

"You only have five minutes," called the bus driver out his small window.

Now the mob tramped the sand to its darkest area. Women scooped handfuls of the granulated blackness and showed it to the camera. "Oh look! It doesn't get my hands dirty," they said.

"Blat–blat," went the bus horn and everyone hurried back to their seats. Tonight they would write on postcards: "Today we saw the black sands beach." I have to smile on these people who thought they saw something but really saw nothing.

A government man once came to Kaimu to tell everyone that building a seawall across the outer bay would protect the beach and stop sand erosion. A few months later Pele got over her fit of laughter, stamped her foot and the whole of Kalapana sank two feet. Government man got the message and never returned with his fancy ideas. Think of all the egg that would be on the faces of the U.S. Corps of Engineers if they had built that breakwater, which today would be under lava.

23 Queen's Bath

That former popular swimming hole known as Queen's Bath in Puna was created by Pele, and now she's destroyed it. Filled it full of lava just like highway workers plug a pothole in the asphalt.

Feel sad on losing such an adventurous beauty spot, but let the feeling be as fleeting as that jungle spa was in existence.

How long had Queen's Bath been with us? Only about as long as today's oldest resident in Kalapana. Before that it was something else.

You see, that crack in the earth filled with water didn't exist 200 years ago. Pele reportedly gave it birth several months before the Reverend William Ellis and a party of New England missionaries became the first outsiders to survey the Puna district in August 1823. The day they ate of imu pig, poi, sweet potatoes, and salt-dried fish under the wind-rattled coconuts at Kaimu black sand beach, they heard residents tell of Pele's antics. She had stamped her feet along the Puna coast. Large long cracks and sink holes had appeared in the district. Several had begun to fill with water from underground springs. In that sense it was a blessing for the natives as pure drinking water often had to be collected and transported from upland forest areas.

One of these sink holes to the south of Kalapana was named Punalu'u. That's because a *heiau* (Hawaiian place of worship) of that name was close by.

During the latter half of the 1800s, that district was grazing land for cattle. Punalu'u pond was their principal watering hole. A good percentage of this herd ended up in Ka'u to replenish those animals killed in the great earthquake, tidal wave, and eruption there in 1868. King Kamehameha V made the gift as land and the animals were his.

That week-long series of earthquakes, as well as others later, began to expand and deepen Punalu'u Pond into the 200-foot-long crack we came to know. But it was never a bath for royalty. Even in 1914 before a road reached the area, it was still Punalu'u pond. Those who detoured to see it were walking the shore trail from Kalapana to view historic Waiula Heiau.

Royalty was swimming in a Puna pool however. This was Warm Springs, just seaward of where Kapoho Village was located. Both the sugar town and the popular springs were covered by lava in 1960. This is the one pond you should cry over as did Hilo businessman Slim Holt, owner of the landmark. His family could remember back around the turn of the century when Hiloans packed a lunch on Sundays and rode the train out to enjoy this beauty spot.

It was sometime in the 1920s when the ever-settling Puna coastline had fallen far enough to make swimming a joy in Punaluʻu Pond. A fresh water spring fed the crack and this floated on sea water tide. Cool swimming on a hot tropical day.

The popular concept is that the swimming hole got its name Queen's Bath from the military stationed in Hilo and those enjoying rest and relaxation at Kīlauea Military Camp. The name started as a popular term and caught on like changing Leʻahi to Diamond Head. I was one of those GIs that took advantage of the now-royalty spa. It wasn't as deep as it was this year. In fact, as

recently as the earthquake of 1975 it gained another foot or more in depth.

But all this should not detract from what Queen's Bath has become to over three generations of young people, my children, and grandchildren included. For those of you who now will never see or experience Queen's Bath, you missed a rare treat. The crack in an ancient lava bed was about 14 feet wide. A swimmer could cover its length in 40 strokes. At high tide it was probably 10 feet deep. We always threatened to take a tape to measure it, but somehow never did. However, the rule was to dive shallow. The water was so clear you would swear it was only three feet deep.

The setting was something out of one of Gauguin's paintings of Tahiti. Clumpy mango trees provided shade. A few ever-swaying coconuts lent atmosphere. Then a thicket of guavas that often hosted passion fruit vines closed in on the water filled crack to give the feeling of a secret Shangri-la in a jungle. Seasonally, when over–ripe guavas, mangos and liliko'i (passion fruit) lay rotting on the ground, you smelled the jungle.

Non-swimmers sat on a low lava ledge on one side of the crack. Here the youngest paddlers could enter the water with ease and get themselves wet a little at a time. Robust swimmers jumped right in, then clambered up the higher ledge on the opposite side. This 10-foot ledge became the springboard for the untamed. A particular delight was to "cannonball," hitting the water with a splash that sprayed the fully-clothed ones watching.

Water still being scarce in Kalapana, residents would often visit Queen's Bath with a cake of soap. This was frowned upon, as well as washing dogs.

The one thoughtless act that diluted the beauty of Queen's Bath for visitors and some Islanders like me was the scattering of lunch trash. It stayed around and multiplied and brought the rats and mongoose. Civic clubs would periodically hold a clean-up day, but the following week the trashing of that beauty spot would begin again.

Now that the place is capped with a frozen bed of molten rock, I'm tempted to say, "See what you've done! Pele couldn't stand the mess and she's covered it all over. It's all your fault."

Someone who should receive praise and never got it during Queen's Bath's lifetime is the Campbell Estate. They owned the area after royalty. All those years they allowed us to use the pool as if it were a public park. No kapu (forbidden) signs. Yet, in a liability sense, they were sitting at the tip end of a dead branch. Mahalo, Campbell Estate.

Well, goodby Punalu'u Pond, alias Queen's Bath. I did shed a tear. A small piece of the fun part of life in Puna has disappeared.

For visitors riding buses or doing it themselves in U-drive cars guided by those throwaway commercial tour guides, this ends their discovery of Puna. Back they go up Route 130.

24 Old Hawai'i and Pele

We will take Route 137 along the Puna coast and experience a Hawaii of yesteryear. The road is paved, but narrow as it offers postcard vistas of the waves crashing on lava ledges, tunnels through tropical rain forests where wild guavas and passion fruit lie rotting beside the road, and reveals hideaway picnic parks. This 15–mile drive is one of the prettiest on the island.

It was in this area that I came face to face with Madame Pele in 1955. The day we met Pele in person we were not supposed to meet anyone.

Please note the plural "we." I was not alone. There are witnesses to the incident. I wouldn't tell you if there hadn't been. For reasons I won't go into, I will not use the real names of my friends. The Pilot, the Scientist, and the Manager will have to suffice.

It was in the spring of 1955. Pele was island building down in the lower Puna area. For a week now a whole section of 'Opihikao had been cut off from the rest of the world by two rivers of lava. Residents that had lived there before were now refugees in a gymnasium in the sugar town of Pāhoa.

The manager of the sugar plantation wanted to inspect what was left of his fields in the abandoned area. He talked the pilot of a small plane to land on the company's crop dusting cinder strip so he could walk the area on foot.

This left two seats open in the plane. The volcano scientist was invited to go because they valued his expertise. I somehow got talked into recording the trip with camera and pen. I am not an overly–brave individual. I'm predictable, reasonably cautious and dependable after a fashion. We landed on the agricultural strip and came to a stop before a wall of trees with the Pilot standing on the brakes and me fingering my buttons as if they were a rosary.

The four of us walked around for an hour inspecting the lava flows and what was left of the sugar fields, a patch of bananas and an abandoned sweet potato farm plot. There were no houses in this section.

Imagine our surprise when we sauntered down a cane field path and came upon a lady sitting at the edge of the road. Now you could say she was a girl, woman or lady. I've checked my notes and I only have her down as a lady. That's the impression I had.

In my estimation a girl would have been under 21. She was older, but not by much. A woman shows maturity in dress and hairdo, a certain domestication of mannerisms, and that look in the eyes that can send children to bed at 8pm. A lady has the charm of sophistication. She carries herself with authority even when sitting. The soft features, clearness of skin, and sculpturing of the nose denotes breeding. This woman had all that at first glance. She wore a red mu'umu'u with black markings that looked like bamboo. She was barefooted. A cloud of jet black hair flowed behind her shoulders.

"Hi," the Manager said, recovering from his surprise.

"Aloha," she answered.

"What are you doing here?" I guess the Manager owned this land so he had a right to ask.

"Just resting in the shade of the sugar cane," the lady said, giving us a radiant smile.

"No one is supposed to be in this area," commented the Scientist. "The National Guard evacuated everyone a week ago. Why did you stay behind? You know you're trapped in between lava flows here."

The lady's smile just grew wider as if that were answer enough.

"What's your name?" I inquired, posing a pencil over my note book.

She said something very musical in Hawaiian that sounded to me like a fern. I wrote the last part down phonetically and it appears as "uulei" in my notes.

The Pilot frowned and turned to us. "I could make two trips and take her out," he said.

"Oh, I won't leave here," said the Lady. "At least not today. I've work to do. Perhaps I'll be ready to go somewhere else next week."

"Well, if you don't want to come away now with us now, you may have to later today or tomorrow," said the Scientist. "We'll have to report you to Civil Defense and they may send a chopper in for you. The eruption has caused an emergency in this part of the island and there are laws to protect people."

"I follow my own laws," said the Lady, and for the first time she stopped smiling.

I remember looking into her eyes at that moment and having the feeling that what I saw I had experienced somewhere in the past. Then I had sudden recall. I had spent a Christmas vacation in a summer home on a frozen lake in northern Wisconsin. We all slept in the living room because the cast-iron stove was there. It was loaded with wood at bedtime, but by dawn it was

freezing to the touch. However, when I lifted the lid to put in more wood, there were two gleaming cherry-red coals nestled in the gray ashes that promised instant rekindling.

The Lady's face was now cold and those same two glowing coals were deep within her gray eyes.

Perhaps my three companions had somewhat the same feeling because the Manager said, "We'll finish our inspection, and if you want to go out with us, you can wait by the plane," and he gestured up the road.

We continued walking. But for some strange reason we had only gone ten feet or so when we all turned around to again look at the Lady.

She was gone!

We ran back. The Manager went into the cane field. The Pilot went up the road. The Scientist jogged down another path. I called. We didn't find her.

A spooky feeling began to creep into all of us like a cloud invading a rain forest at dusk.

"I think it's time to go," said the Pilot. No one disagreed.

We entered the plane and taxied to the end of the cinder strip. The Pilot gunned the engine and stood on the brakes as long as he could. We lurched forward. Halfway down the field it was obvious we were too heavy to clear the trees ahead. Landing on a small strip was one thing. Taking off with a full load was another.

We stopped. "Someone's got to get out," said the Pilot. "I'll come back for you later."

That "you" was directed at me. I was sitting next to the door. I made no protest and got out of the airplane.

I was scared. Not in the 'fraidy cat-way, but apprehensive, to say the least. I did not want to socialize at this time, and as the underpowered airplane diminished to a dot, then disappeared, I dreaded social contact more and more. My thoughts raced around Pele sightings I had heard. The common one is that she's hitchhiking, gets into a car, asks for a cigarette, lights it with the tip of her finger, then when the driver looks the other way, she disappears. In all modern Pele legends she has this disappearing act down pat.

An hour dragged by and when no lady appeared I began to laugh at myself. That wasn't Pele. She was real flesh and blood and was probably a resident who had just hidden from being evacuated. I had a name and description. I would solve this incident once I got out.

The Pilot in that beautiful little airplane did return and an hour later I walked into Civil Defense headquarters and the refugee gym. The Pilot, the Scientist, and the Manager were there. The four of us grilled the residents of Puna. The lady's name and description didn't fit anyone. The National Guard commander and Civil Defense workers all assured us they had done a thorough job in getting everyone out—people and pets.

Now you know why I'm continually walking the trails of Kīlauea and why I wrote this book. I don't walk alone; a lady accompanies me. She walks barefoot.

25 A Special Day with Pele

Of all the dances and walks I have had with Pele, the one on March 14, 1955, is outstanding. This event made headlines, and pictures of it took up the entire front page of Hawaii newspapers, made *The New York Times*, and the history books.

Again, I was not alone with Pele. There were five of us: Dr. Gordon Macdonald, Kīlauea volcanologist, Seismologist Dr. Jerry Eaton, of the United States Geological Service, Curtis Kamai of the Hawaii Volcano Observatory, *Life Magazine* photographer Nat Farbman, and myself. We were all riding in a jeep on Route 130 in Puna with a portable seismograph searching for earthquake signs that Pele makes when she's dancing under the earth's surface. An eruption in the area was a good possibility, according to the scientists. Lunchtime came, and we stopped to eat sandwiches along side the road. Before us was a plowed field where farmer Masayuki Nii had planted cucumbers.

1:47pm – The earth began to shake. Small cracks about a half inch wide and only a few inches deep began splitting up the field in front of us.

1:48 – The loose earth began to undulate. It shifted back and forth, moving like a thing alive. Dr. Macdonald runs into the field with his 16-mm movie camera and begins filming.

1:50 – Loose patches of earth began to heave and fall, rising and falling as if giant ground hogs were at work. The cracks in the earth began widening. Dr. Eaton approaches a crack and prepares to lower a thermometer on a thin wire into the vent.

1:52 –A thin veil of milk-white smoke issues from the cracks. Dr. Macdonald stands over one crack trying to see what might happen next.

1:53 – The smoke gets thicker and begins to billow out of the cracks. We all smell sulphur. Nat Farbman and I are shooting film like there is no tomorrow. I have double duty to keep checking my watch and jotting notes in a pad because I also have to write a story for my newspaper *The Honolulu Advertiser*.

1:54 – A hissing sound is heard much like when a cat spits at a dog. The scientists, who have been within inches of the cracks, now move back.

1:56 – The sides of the cracks in the earth turn purple with the heat. The sulphurous white cloud is now shooting 10 feet into the air. The hissing sound increases to where it's like a trailer truck letting off air brakes.

Are you, the reader, aware of how fast it's happening? This is not a waltz. This is a fast tango with Pele.

1:57 – A bubble of lava, molten and dark red, squeezes up through the cracks and pours over the ground. It pops when it hits the moist earth, gas escapes and chunks of lava are thrown into the air. Dr. Eaton calls it a "baby fountain."

1:58 – This minor fountain action now throws lava more than two feet into the air. Again, humans retreat.

1:55 – There is an audible rumble from one of largest cracks. We can feel heat. The sulphur smoke increases and blows into our faces bringing tears and choking. Dr. Macdonald warns us that if we get a mouthful of the stuff and can't breathe, to lie down and scoop away a patch of earth. Lean into the depression and get a breath of fresh air, then get up and run like hell.

1:59 – Lava now comes out of the ground like a pulsating, gasping monster. It rises and falls 8 to 10 feet into the air. Lava runs over the ground. Again, we move backwards, this time toward our jeep. A grand escape plan is beginning to make sense.

2pm – The thing is now out of hand. Pele is beginning to show off. The fountains are rising. Lava is flowing, rasping its way over the ground like a reptile.

2:01 – The many small fountains now consolidate into one. A cone of harden lava is beginning to build where the molten rock falls back on itself in the vent area. Small rivers of lava began to find their way downhill from the field. The action in front of us is well over 25 feet high.

We regroup on the highway a safe distance away. Dr. Macdonald tells us, "we have just seen a volcano born." He knows such a scene has never been photographed before or maybe studied in as minute detail as we have done here. Dr. Eaton is a little more specific. He says, "This is the first time we know of that anyone has recorded in such detail the birth of a volcanic vent." Up to now, only Italian peasants and a Mexican farmer have ever seen anything like what happened in this cucumber patch, and no photographers were present at those events.

1955 The birth of a volcanic vent is recorded in person, in detail.

26 More of Old Hawai'i

Our tour along Route 137 begins over the reforested lava flow of 1750 that created the scenic black sand beaches at Kalapana. Beyond that, you enter the tropical rainforest area where greenery presses in on you. Here and there vacation homes are tucked away on large lots.

A black sand beach at a spot called Kehena is located down a 50-foot cliff. It has a reputation for being a "clothes optional beach."

A new vacation subdivision further on was created on a 1955 flow where the developer simply bulldozed the surface level and today people are building homes.

The bulldozing had a historic effect. This was the first time someone, who had previously owned waterfront property, had his land extended seaward by a flow. That person figured he had a bonanza of additional free land, and he began to level it for development. The State of Hawaii said they owned any new land built by lava. The case went all the way to the Supreme Court, which ruled in favor of the State.

Now this brings up a question concerning the lava flow from Pu'u O'o that started in 1984 and is still active today. Much of this flow has run down Kīlauea's eastern flank over National Park land, but it has also built several miles worth of new land out into the ocean beyond the former edge of the island. The National Park System is acting as if it owns this new land. So far, the State of Hawaii has been silent, perhaps because no one has made an issue of it.

There's not much left of 'Opihikao Village which was the major settlement along this coast before 1900. Today there is a re-built church and several homes. A house on your left, just past the junction tucked away in banana trees and colorful tropical foliage, is typical of older Hawai'i and a favorite with photographers and artists.

There's a story about this place I'd like to share with you. Back in the 1800s, villages such as 'Opihikao were serviced by coastal sailing packets that delivered people and goods to rural areas along the coast. This colorful transportation disappeared after 1900 when automobile roads were established.

Sometime around 1880 or so, a young farmer from Japan arrived with his worldly goods in a wicker basket. He went inland and established a small farm where he grew dryland taro, sweet potatoes, bananas, and other crops. He never left the district, which isn't strange because even in 1946 when I became involved in this story, a few children in this village had never even been to Hilo.

One day in 1946 the police were notified that an old Japanese man had been sitting on the lava shore for a week fronting the village. When they questioned him, the farmer said he was waiting for the sailing packet because he was returning to Japan to die. He had all his worldly goods, including his life's savings, in a wicker basket. The old gentleman was taken to Hilo where they bought him an airplane ticket to Tokyo.

MacKenzie State Park is worth a look and a walk about. This park has its good and bad side. On the good side, a section of the old King's Highway, a horse road built by convict labor under Kings Kamehameha III and IV back between 1850 and 1860, runs through this park. (Better examples of this road are in north Kona.) The park is forested with ironwood trees and is a favorite place for weekend fishermen casting off the cliffs. Walk along the waterfront for an exciting picture. Check the style of picnic tables.

On the bad side, the park will appear abandoned by the State. The water tank and drinking fountain are gone. The toilet facilities are barely usable.

Isaac Hale County Park at Pohoiki is worth another stop. This park with a boat launching ramp has become a haven for fishermen. There will be surfers out if the waves are up. Walk past the caretaker's house to the head

A view of Puna's coast.

of the bay and then along a short path into the jungle. You'll find a fresh water tidal pond heated by volcanic action.

As you continue along the road for the next two miles, take note of the nature of the road, the scenery, foliage, a small coconut grove and the types of houses. This is the closest Hawaii comes to resembling rural Tahiti.

Ahalanui Warm Springs In the middle of all this is a home in a coconut tree grove that was once a private estate. It is now owned by Hawai'i County and has been turned into a park called Ahalanui. At oceanside is a pond that is heated by volcanic warm springs. The feature was once a fishpond so has an opening to the ocean. At low tide the pond is bathtub warm. At high tide when the waves come crashing into the pond, the water is less warm. The County Parks Department has refashioned the pond for safe swimming, and a life guard is on duty much of the time. Nothing could be more scenic or delightful than this spot.

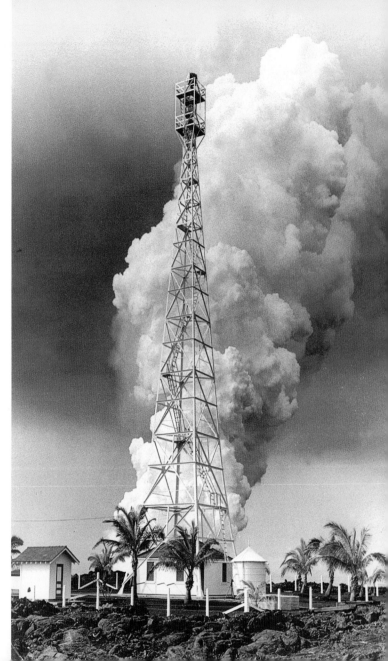

The lighthouse at Cape Kumakahi was saved from Pele by earth dikes built at an angle to the flow.

27 Cape Kumukahi

The cinder road at the junction of routes 127 and 132 leads to the easternmost point of this island, and a weird scene that has been blown up into a legend. It's time to put the record straight.

Since this is also the easternmost point of our state, it was an important site for a lighthouse. Up until the eruption of 1960 that destroyed the nearby town of Kapoho, this was a manned beacon with a lighthouse keeper. After the lava covered the town, it flowed down this road and did away with the lighthouse keeper's home and maintenance buildings. Then as the river of molten rock knocked at the door of the lighthouse, it suddenly flowed around the light tower to enter the sea beyond.

A modern made up legend has it that Pele spared the lighthouse because Pele was in love with the light house keeper. Well, Pele had plenty of help in getting the flow to do what it did at the light station. We can thank Dr. Gordon MacDonald, former director of the Kīlauea Observatory, and a host of state workers on bulldozers for what really happened.

Dr. MacDonald had the theory that a lava flow could be diverted from its path under certain circumstances considering the topographical lay of the landscape. He figured an earth dike built at an angle to the flow could force the molten rock to head in a new direction.

The first dike of earth built to keep the flow from covering Kapoho failed as the lava simply went under the dike because it was made mainly of loose soil. The second dike built to save the area's elementary school located at the junction of 130 and 137 failed because the tilt of the land seaward was not great enough.

A third dike was pushed into place by a swarm of bulldozers to save the lighthouse. This was a partial success. Lava overflowed the dike, but slowed enough to begin cooling at the edges fronting the lighthouse tower, and this hard rock front provided enough of a natural barrier so the flow moved on around the structure and toward the ocean.

28 Fishing with Pele

When Pele takes a bath, she does it with class. Although the Pacific Ocean is her bathtub, she heats the water. I'd like to see us mere mortals try to do that sometime!

Six times over the past 40 years I've sat on the edge of Pele's tub while she frolicked in the water. She's dynamic, yet modest as she hides in a salty steam cloud that billows thousands of feet into the sky.

When she plays in the water, it isn't with a rubber ducky. She cooks fish and eels, parboils seaweed, steams opihi clinging to rocks, and then delights in cooking the rocks until they turn a bright pink. Oh, by the way, Pele uses a soap that colors the sea a pea-green.

This is also a noisy bath with lots of hissing, gurgling and base drum booming going on. The latter is from hand grenade-type explosions that form sprays of fine black sand when gas-laden molten lava hits the water.

Sometimes Pele is coy about getting her feet wet as if its's not quite bath time. She'll sit on the edge of her tub and gently dip her fingers into the cool Pacific. Dip –hiss, dip–hiss, she plays at the water's edge with liquid rock stretched out like taffy that dribbles into the water.

One time I jumped right in and took a bath with Pele. In fact, two of us went swimming in her tub. Her invitation and bath water were irresistible. We couldn't help ourselves.

It was during the 1960 eruption when lava was extending the eastern point of Cape Kumukahi a half-mile seaward. My fishing buddy, Allen Chang, and I were eavesdropping on Pele's bath when we noticed a large school of fish had become trapped in a bay. Pele was hidden in her ocean steam and her overflowing tub had sent a river of hot water across the mouth of the small bay, corralling a school of fish within.

The fish were *moi*, a great island delicacy of the goatfish family. They crowded the bay like cattle being herded into a box canyon.

"Whatdaya think?" asked Allen. "Is the water too hot?"

I took my shoes off and tested it. "Warm." "Let's go," he said, beginning to strip down. In minutes we were swimming with the fish. You couldn't stroke in any direction without hitting them. Fish were under our armpits and between our legs. Silvery-gray wiggling forms ranging from two to five pounds apiece.

The ocean was getting noticeably warmer by the minute. Pele was only a few hundred feet away splashing, foaming, sputtering, blasting curtains of black sand, and generally creating a beautiful spectacle.

So we went fishing. Treading water, we would feel around for the biggest, fattest *moi*. We quickly became experts at petting fish. The prize ones would be cradled in our arms and then thrown ashore to flop their lives out on the rocks.

About 15 minutes of this and we had to call a halt. We had as many fish as two men could carry, when you consider we had nothing to carry them away in. Fish littered the shore like driftwood after a storm.

In the end, we tied the legs of our pants and the sleeves of our shirts together, and filled our clothes with fish. Then we walked the mile to the car in our underpants. We could hear Pele laughing every step of the way.

That evening we talked a restaurant owner in the town of Pāhoa into cooking our fish. We went around the community inviting everyone to the free meal.

The author and childhood friend Allen Chang with enough hand-caught moi to feed a village, thanks to Pele's antics at the beach when she cut off an entire school of fish from the open sea. Chang and Morse waded into the warm water to take advantage of its stunning effect on the fish.

Fountaining in Kapoho in the 1960s.

29 Kapoho Village

Please stop your car at the junction of Route 37 and 132 and look around. Your imagination will have to come into play here as you try to visualize a town about the size of Pāhoa that extended from this junction for a half mile up Route 132.

To your immediate right was an elementary and intermediate school. Bordering the road were homes, a post office, stores, gas station, a theater, and so forth.

The cinder cone you see to the right as you look up Route 132 was the eruption source in the spring of 1960. Within 31 days, Pele covered this entire area with lava and extended the eastern point of Hawai'i further toward the sunrise.

The first hint of what might happen to Kapoho was a string of earthquakes that set dogs howling in the night. By dawn a series of cracks criss-crossed the paved road through the village. Then a large crack opened in a field to one side of the business buildings. From this issued a curtain of steam, followed by a curtain of molten magma that produced lava ash that fell over the town.

Men climbed upon roofs to shovel the warm powdery material off before the buildings collapsed. Other owners called in flat bed trucks, jacked their building up, loaded them on the trucks and took them away.

In short order the entire village was evacuated except for a few policemen, State civil defense workers, newspaper reporters, and photographers who hung around to record the end.

NEW RIFTS RIP PUNA

Exclusive Story, Photos Show Birth of Volcano

History Made at Site Of 2nd Sunday Eruption

(EDITOR'S NOTE—Five men made volcanic history Sunday in a cucumber patch at the edge of the Pahoa-Kalapana road. They were the first men to see and follow the emergence of molten lava from soil...

The destruction of Kapoho Village began with earthquakes shaking the community and streets that split in two on January 13, 1960. By the end of the month it was all over for residents of what had been a peaceful country community.

The youthful principal of Kapoho School poses for posterity with an item she was able to rescue during the destruction of Kapoho Village by an eruption of Kīlauea Iki in January 1960.

By GORDON MORSE

ON THE KALAPANA ROAD, Hawaii, March 13
(By radiophone)—I watched the birth of a volcano today. I stood with four other men while the earth ripped

Pele wasn't finished
in Puna in 1960, but
continued to assert her
presence by by spewing
forth molten lava and
belching sulphurous
smoke that blocked roads,
singed 'ōhi'a forests,
and threatened to scorch
houses in her path.

30 Aloha to Warm Springs

Besides losing the town of Kapoho, residents of the Big Island also lost a spot called Warm Springs located on the outskirts and now under 50 feet of lava. Only two humans attended the burial of this historic landmark.

Warm Springs was the true Queen's Bath of Puna. Its tradition as a spa went way back. It was a large deep pool of warm Saturday night bath water. Only one side of the pool was open to the public. It had a park-like atmosphere of picnic tables, manicured flower beds, and orchids cascading from tree limbs. The pond was backed by a cliff of ancient lava overgrown by trees that played host to veils of vines and umbrellas of ferns.

All this shade rendered the volcanic-warmed waters a black satin mirror. Bathers were transported into dreams of what they imagined a South Sea Island scene should be: the hideaway pool in a tropical jungle with orchids floating like soap suds on the surface. One had the feeling it could be Ponce de Leon's bathtub to restore youth. It did make you feel young at heart.

This paradise spot was owned by Slim Holt, a Big Island pioneer in post–World War II tourism. He lived in Puna and understood Pele, but one night in 1960, he found it hard to forgive her for destroying his treasure. Twenty years later he went to his grave still wondering why she paved over Warm Springs.

Toward evening of that fateful day, island photographer Robert Wenkam and I sat alongside the road leading to Warm Springs discussing the situation.

We had just been barred from the area by police. A wide 'a'ā flow was headed for the ocean beyond as it bulldozed through papaya groves and virgin forest.

"It would be a shame to have such a historic beauty spot destroyed and no record of it on film," said Wenkam.

"My thoughts exactly," I said. "Do you think we can get around the police and Civil Defense blockade without being detected?"

"Worth a try," said Wenkam. "It will entail some walking. The authorities will be on the lookout for our car."

The walk was mainly along an abandoned railroad right-of-way, then through a papaya patch to the edge of the jungle. From here it was a good guess as to the direction of Warm Springs about a mile away. It was an odd night, lit a fuzzy blood-red by the glow of Pele's 500–foot–high fountain about two miles away.

"Looks like we can get to the pond, but getting back out might be a problem," I said. "If we stay until the area is covered, then the lava flow will be upon us and change the looks of the area and our retreat will be hectic. We could get lost and be in deep trouble."

For security on the retreat we tore our handkerchiefs and our T-shirts into strips. These we tied on branches to mark our trail as we proceeded through the jungle.

We arrived at Warm Springs about 9 o'clock. The flow was still about 1000 yards away and moving maybe ten yards a minute. It only took a few moments for me to begin feeling that we were the last people on earth, about to witness the destruction of the world. Before us and to our right was a 15–foot wall of molten rock relentlessly bulldozing its way across the land. To our back was the ocean. The way to escape our soon–to–be ravished paradise was our trail of torn cloth through a forest.

We stripped and went swimming. We laughed and splashed in moonlight dyed red by the glow. We picked orchids and floated them on the water making believe they were treasure ships. Mine was white with a yellow sail. Wenkam's was lavender with a white sail.

"All the fiction of *Fantasy Island* will not beat this night," I told my swimming companion.

Then Pele invaded our spa. She came flaming through the forest like a monstrous bulldozer. We dressed quickly before she could laugh at our nakedness, took pictures of her coming across the park as she picked gingers and heliconia and ate picnic tables. Finally we

Warm Springs in 1960 as lava filled it. The author and renowned photographer Robert Wenkam took a last daredevil swim here before "Pele invaded our spa."

had to scamper up the cliff back of the warm pond. There we sat, each using up a whole roll of film as the lava surrounded Warm Springs, tilted the trees with their orchids into the water and filled in the pond with molten rock. Flames reflected over the dark waters as the lava hissed into its depths. We left on the run when steam choked the forest around us.

Going uphill on Route 132 there will be a vanda orchid farm on your right, and then papaya groves on both sides of the highway. At the top of a rise you will enter a forested area with Lava Tree State Park on your right. This is worth a walk, especially if flowering plants are in bloom. There are mosquitos. Then back to Pāhoa.

Pele's Travels

Our sightseeing and experiencing a volcano may be completed, but somehow Pele never fades from our consciousness. That's because the goddess is not a homebody. She travels. I've seen her, in of all places, at the mouth of the Columbia River in Oregon.

For years now, a weird story has circulated about a sand flow there that resembles the Big Island's lava flows from Kīlauea or Mauna Loa. They say that a pattern of sand flowing down a dune the way lava flows down the side of a volcano, is a reminder that Kenue is buried in that far-away land. Kenue was reported to be the first foreigner to be buried with ceremony in that part of the Northwest. That's debatable, but we won't let that fact dilute our story.

Historians do know that Kenue left Waikīkī on February 26, 1811, with 23 other Hawaiians aboard the sailing ship *Tonquin* bound for the Northwest. The Islanders had been recruited by the North West Fur Company to help build Fort Astoria, the first settlement in that part of the world. They were also to aid in fur trading, and to augment the crew of the *Tonquin*.

King Kamehameha personally okayed the hiring. Each Hawaiian had a three-year contract for which he received his clothing, food, and $100 in merchandise from the company storeroom.

The adventurers reached the mouth of the Columbia on March 22, but couldn't go upriver as a brisk Northwest wind was blowing and heavy surf made it impossible to determine the river's channel.

Two attempts to scout a safe passage in small boats were made over three days by the *Tonquin's* crew. All ended in failure resulting in the disappearance of seven crewmen.

On the third attempt, Kenue and another Hawaiian were recruited to row the boat along with three other crew members. Their boat too, was capsized by a wave. Two of the sailors disappeared in the foaming sea and fierce rip–tides of current where river met the ocean. The Hawaiians and the remaining sailor stripped off their clothing and managed to swim to the swamped boat. They turned it over, bailed it out and made for shore.

Night and the cold set in and the Sandwich Islanders gave up rowing and huddled together for warmth. Toward midnight Kenue died of exposure.

Morning found the shivering seaman and the surviving Hawaiian making a safe landing on a beach. There they placed Kenue's body in a tree to keep it out of reach of animals. The sailor went down the shore to look for the *Tonquin* while the Hawaiian stayed with the beached boat as he was too weak to walk.

Meanwhile, the *Tonquin* had finally made it through the river's entrance and the two survivors were found.

Late that afternoon Kenue was buried in the sand dune by his countrymen. He was laid to rest with a biscuit under one arm, a piece of pork under the other, and a wad of tobacco under his chin. One of the Hawaiians acted as the *kahuna* (priest). He filled his hat with sea water and sprinkled it over the body and all who attended the ceremony. After a chanting ritual, they filled in the grave with rocks and sand.

All of the above was recorded by Washington Irving in his book *Astoria*, a history compiled from fur company reports, and by Gabriel Franchere, a fur trader from Montreal who was on the scene at the time.

It wasn't many years after Kenue was buried that beach walkers in the vicinity of the Columbia's entrance noticed a weird pattern in the sand. It was as if the sand was flowing in streaks or ribbon-like sculptures. Interesting. But no one paid much attention to the phenomenon.

In 1976, after many Hawai'i residents became more aware of

Hawai'i's migrating natives and their contribution to early settlements in America's and Canada's northwest, the sand patterns took on a new meaning. Islanders now exploring this beach area took one look and recognized the patterns instantly.

There was no doubt about it. The forms were sand images of lava flows.

It's true.

I've seen it too.

And so the legend has begun to develop. Goddess of volcanoes Pele reached far afield to create a reminder that Kenue is buried beneath those sands.

Pele's far-reaching handiwork? Lava flow-shaped sand dunes at the mouth of the Columbia River in Oregon seem to suggest it.

Acknowledgements and Credits

Additional Art and Photography Credits
Joann Morse, hand-tinting of photograph, p. 2
Buzz Belknap, pp. 14-15, 31, 36, 42, 49
Eric Nishibayashi, p. 56
Jodi Belknap, p. 53
Reginald Ho, p. 93
Gordon Morse *Honolulu Advertiser* photos, pp. 86, 87, 90.

Cartography Credits
C. Christina Heliker, of the USGS Hawaiian Volcano
Observatory, provided up-to-date lava flow information
(as of February 2005) for the maps on pp. 8-9, 61, and 75.

Notes on Production Technology
This book was designed and laid out using MacIntosh
computer equipment (G5s and G4s) and an Epson
scanner and printer. Liason between the author/
photographer and the design and production house which
are located on two separate Hawaiian Islands, was greatly
enhanced by the use of Apple MacIntosh iDisk service
and pdfs. Adobe CS® was the design and publishing
program used.

Typography
The primary typeface used in this book is GoudyCI,
based on a classic font originally designed by the
renowned 20th century typographer and book designer/
publisher Frederic Goudy. It has been technically
enhanced by Coconut Info® of Honolulu to allow easy
use of Hawaiian language pronunciation marks—the
ʻokina (glottal stop denoting a halting of breath before a
vowel), and kahakō (macron, or stressed vowel).

Notes